**Advanced Topics in Science and Technology in China**

Volume 59

Zhejiang University is one of the leading universities in China. In Advanced Topics in Science and Technology in China, Zhejiang University Press and Springer jointly publish monographs by Chinese scholars and professors, as well as invited authors and editors from abroad who are outstanding experts and scholars in their fields. This series will be of interest to researchers, lecturers, and graduate students alike.

Advanced Topics in Science and Technology in China aims to present the latest and most cutting-edge theories, techniques, and methodologies in various research areas in China. It covers all disciplines in the fields of natural science and technology, including but not limited to, computer science, materials science, the life sciences, engineering, environmental sciences, mathematics, and physics.

If you are interested in publishing your book in the series, please contact:

Dr. Mengchu Huang, E-mail: mengchu.huang@springer.com

This book series is index by the SCOPUS database.

More information about this series at http://www.springer.com/series/7887

Hanhua Zhu · Lei Shi

# Methodology of Highway Engineering Structural Design and Construction

ZHEJIANG UNIVERSITY PRESS
浙江大学出版社

Springer

Hanhua Zhu
Highway Bureau of Zhejiang Province
Hangzhou, Zhejiang, China

Lei Shi
Tibet Lingzhicao Construction Consulting
Group Co., Ltd.
Lhasa, Xizang, China

ISSN 1995-6819 ISSN 1995-6827 (electronic)
Advanced Topics in Science and Technology in China
ISBN 978-981-15-6543-4 ISBN 978-981-15-6544-1 (eBook)
https://doi.org/10.1007/978-981-15-6544-1

Jointly published with Zhejiang University Press
The print edition is not for sale in China Mainland. Customers from China Mainland please order the print book from: Zhejiang University Press.



This Springer imprint is published by the registered company Springer Nature Singapore Pte Ltd.
The registered company address is: 152 Beach Road, #21-01/04 Gateway East, Singapore 189721, Singapore

# Foreword

This is a book about highway engineering structural design and construction technology, which is very professional, and is the authors' summary of decades of practical experience and his understanding of theoretical thinking. The author Zhu Hanhua is a senior engineer with rich experience in engineering design, and he has accumulated a large number of cases in years of practice on highways in Sichuan and Tibet, where the design and construction techniques are very difficult; author Shi Lei, graduated from Zhejiang University, has been building roads and bridges in Sichuan and Tibet for more than 20 years, and has rich experience in construction and unique theoretical insight. The joint work of both of them, with theory and practice, problem-centered, targeted and down-to-earth, can be a practical manual for technical personnel in highway engineering structural design and construction.

To tell the truth, I don't understand most of the contents of this book. I don't understand engineering technology, let alone the design and construction of highway engineering structures. The reason why I am writing the preface to this book is that I sincerely admire the two authors' spirit of consciously using philosophical methodology to solve problems of engineering technology, and I want to support and encourage this exploration. The two authors, one from Comrade Mao Zedong's hometown and the other from Comrade Deng Xiaoping's hometown, were full of reverence for these two great men from an early age. They liked reading their books and learning their spiritual style and ways of life when they were young. They read through the *Selected Works of Mao Zedong* (Volume 1–4) and the *Selected Works of Deng Xiaoping* (Volume 1–3), especially inspired by the world outlook and methodology of the two great men in solving major problems in China's revolution, construction and reform by applying Marxist philosophy. They get inspiration and reference from it, and want to use the Marxist world outlook and methodology to solve the problems in their work, that is, the design and construction of highway engineering structures.

The authors of this book, looking at the similarities and differences between the two modes of thinking in Chinese and Western philosophy and their complementation, think that in the field of engineering technology, the advantages of the two modes of thinking should also be absorbed to avoid their shortcomings. For example, Western philosophy pays attention to analysis and synthesis, induction and deduction, while Chinese philosophy pays attention to integration and connection, intuition and understanding, etc., which can all be applied to highway engineering structural design and construction. The world outlook and methodology of Marxist philosophy are more instructive. For example, both "two-point theory" and "key theory" are required; not only to grasp the main contradictions and major aspects of those contradictions, but also to cover the whole to achieve a good overall balance. The authors believe that simplification of complex problems is an important method for mechanical analysis of engineering problems, but to realize that, the unity of overall control and the control of details should be achieved. Simplification of complex problems is conditional, and maintaining a stable equilibrium of the system is the key.

The authors of this book also pay attention to absorbing spiritual nutrients from the red culture. For example, they extracted a method of comprehensive analysis from Comrade Mao Zedong's *On Protracted War*, considered that the "three magic weapons" of the Communist Party of China, that is the united front, armed struggle and party building, can be used by analogy and reference in the structural design and construction of highway projects; Comrade Deng Xiaoping's "emancipating the mind and seeking truth from facts" can also become the ideological line followed in engineering design, etc.

What is commendable is that the fifth chapter, "Application of Typical Engineering", and the sixth chapter, "Analysis of Typical Problems", of the book analyze a large number of problems and cases in the structural design and construction of highway engineering, and contain the authors' unique line of thought and suggestions, which fully embody the "living soul" of the dialectics of "circumstances alter cases" and "everything depends on time and place". Only the colleagues engaged in structural design and construction of highway engineering can realize the mystery and benefit from it.

The last part of the book includes Academician Pan Jiazheng's article *Philosophical Thinking in the Construction of Water Conservation Projects* and my article *Dialectics of Reform and Opening-up*. I think the authors mean to use these two articles as evidence to illustrate the fact that large-scale projects cannot work without philosophical thinking and guidance. Academician Pan's article explains that the construction of water conservation projects should be guided by a philosophical world outlook and methodology. My article explains that the reform and opening-up is also a big project and cannot be separated from the guidance of Marxist dialectical materialism and historical materialism. This article of mine is my experience of learning from General Secretary Xi Jinping's important statement on comprehensively deepening the reform. General Secretary Xi Jinping has set an example for us in the process of leading the comprehensive deepening of the reform, and he is a model of learning philosophy and using philosophy. Xi Jinping's Thought on Socialism with Chinese Characteristics for a New Era is the latest development of Marxism, and is Marxism in the twenty-first century.

The inspiration from my reading of this book is that a Marxist world outlook and methodology can not only guide the governance of the country, but also guide engineering technology. Engineering and technical personnel should not only have excellent professional qualities, but they should also have a basic philosophical background. Using a correct world outlook and methodology as guidance, they will often get more with less and achieve greater success.

Once again, I would like to pay tribute to the two engineers and technicians for their valuable philosophical exploration.

Beijing, China Feng Jun

**Feng Jun** is a philosopher, professor and doctoral supervisor, who has successively held the following positions: Vice President of Renmin University of China and Dean of the School of Philosophy, Chairman of the Chinese Society of the History of Foreign Philosophy, Deputy Chairman of the Philosophy Teaching Steering Committee of the Ministry of Education, Executive Vice President of the China Executive Leadership Academy in Pudong, Deputy Director of the Central Party History Research Office, and Executive Vice President of the Chinese Communist Party Historical Society, etc.

# Preface

Comprehensive statistical data show that among the death toll of various incidents of collapse in civil engineering, underground works account for 32.6%, foundation pit excavation and retaining wall collapse account for 23.9%, the collapse of temporary facilities and supporting elements account for 32.6%, and structural collapse such as road bridges account for 9.9%, but the number of critically damaged bridges with a service life of less than 30 years accounts for 64% of the total number of critically damaged bridges, even though structural defects such as road bridges in developed countries in Europe and the United States also account for 11%. The root cause of the above high incidence of engineering damage is the problem of metastable equilibrium of traffic engineering structures. In addition to the engineering environmental conditions and human factors, the existing theory of equilibrium and stability of engineering structures has implicit "deformation coordination" restrictions in its application. However, when some engineering structures are stressed, their material properties, microstructure, constitutive relationship, integration and force transfer path will change, which does not meet the implicit constraints of deformation coordination theory. If we regard "deformation coordination" as a theoretical problem and need to innovate the theory and rebuild the system, we will complicate the engineering problem and have limited effect in solving the actual engineering problem. Although it is complicated to deduce, the essence of solving the problem is still the control method. If "deformation coordination" is regarded as the problem of control method, as long as the technology and structure are innovated, the engineering problem will be simplified, and the effect of solving the actual engineering problem by using the existing theory of equilibrium and stability of engineering structures is better. Both of them have the same goal of solving the problem, but the process and method are different, that is to say, the state of stress and deformation of the control engineering structure is different. The former complicates the simple problem while the latter simplifies the complex problem. From the perspective of methodology and solving the practical engineering problems, it is more appropriate to regard "deformation coordination" as the problem of control method. In view of unfavorable structures or composite engineering structures such as flexible bending members, broken surrounding rocks, soft soil foundation, etc., the application of engineering mechanics to solve structural problems of engineering should be studied on the premise of following methodology, and the problems that may occur in actual engineering should be exposed to the research and test links, so as to solve them in the design and construction links, avoid problems of quality and even of safety in actual engineering use, and ensure the safety of people's lives and property.

Chairman Mao Zedong pointed out the construction of Beijing Metro Line 1: "careful in design, careful in construction, in the construction process there will be a lot of mistakes, failures, we should always pay attention to correcting them", which is an example of using philosophy to guide engineering; Comrade Deng Xiaoping's guidance on reform and opening-up calls for "emancipating the mind and seeking truth from facts", which is both an ideological line and dialectics. Both Mao Zedong and Deng Xiaoping were masters of "learning war in war". Whether it is war or revolution, construction or reform, they cannot be separated from materialist dialectics. Inventions in science and technology need to learn and master materialist dialectics. The greatest enemy of innovation is becoming accustomed to it. Technological progress and construction complement each other. It is difficult to develop technology without construction projects.

Inspired by the article *Philosophical Thinking in the Construction of Water Conservation Projects* of Pan Jiazheng and guided by Feng Jun, professor of philosophy, the authors of this book try to discuss the practical experience of structural design and construction technology of highway engineering from the perspective of methodology. It is divided into three parts: the methodology of law and theory, the method of the analysis of comprehensive research, the simplification of complex problems, and the application of typical engineering and the suggestion of problems. At the end, there are guides for facilitating the personnel of engineering construction in using philosophical methods to guide engineering construction, but not limited to philosophy to guide engineering discussion.

This book is the authors' attempt at writing about the problem of the methodology for the construction of highway engineering structural design, there will inevitably appear various defects and mistakes, and all readers are welcome to criticize and correct!

Hangzhou, China Hanhua Zhu
Lhasa, China Lei Shi

# Contents

# Chapter 1
# Philosophy and Thinking Methods

From the 8th century B.C. to the 2nd century B.C., the wisest men of all mankind were born almost at the same time, for example, Confucius, Sakyamuni, Socrates, Plato and Aristotle, who made this period the so called "Axis Age". Ancient Greek philosophers mainly considered the relationship between human beings and things (the objective world). Philosophy and science were not separated and natural philosophy developed. Ancient Indian philosophers mainly considered the relationship between human beings and god (transcendental world) and transcended the world of man's experience. Ancient Chinese philosophers considered not only the relationship between human beings, but also that between human beings and nature (the relationship between nature and human beings), emphasizing harmony and a win-win situation. The fusion of Chinese culture (changing the unsuitable part, absorbing the advanced part from the outside and strengthening the adaptive ability) absorbs the advanced culture from outside on the basis of adhering to Chinese culture, and constructs a more advanced culture to serve the people, just like adding an alloy into steel to forge a better steel alloy. For example, every bit of progress in Chinese history can be seen in the course of cultural integration, such as the Qin, Han, Tang and Yuan dynasties, which combined various cultures, especially the promotion of harmony among Confucianism, Buddhism, Taoism and so on, and the contemporary integration of Marxism and Western science and technology has achieved inclusiveness in both Eastern and Western civilizations and ways of thinking, but the Ming and Qing dynasties did not grasp the general trend and they lag behind the pace of the Western Industrial Revolution, that is, reform and opening-up and grasping the general trend will lead to development, while conservative, closed and misjudgment of the era will lead to backwardness. Therefore, the British scholar Joseph Needham's (1900–1995) question "Although ancient China made many important contributions to the development of human science and technology, why didn't science and the Industrial Revolution take place in modern China?" can be easily understood, as long as the state adheres to the state policy of "the combination of reform and opening-up, grasping the general situation, centralized planning and overall promotion", the key



H. Zhu and L. Shi, *Methodology of Highway Engineering Structural Design and Construction*, Advanced Topics in Science and Technology in China 59,
https://doi.org/10.1007/978-981-15-6544-1_1

is to grasp the general situation and the united front and the Party's construction, and there will be continuous progress and the problem will be solved.

The ancient Greek and Mediterranean civilizations had the convenience of navigation and were located at the junction of the east and west world, with relatively developed commerce, while China was mainly a farming civilization with relatively developed agriculture. Western culture is a confluence of three cultures: 1. The Ancient Greek and Roman cultures; 2. The Hebrew culture of Judaism and Christianity; 3. The Arab culture. After the Renaissance, Western countries paid more attention to scientific research and technological innovation, especially in the United States, where the influence of Protestant Christianity was great. The characteristic of the methodology of Western philosophy is to pay attention to logical thinking. It stresses not only confirmation but also falsification, not only the necessity of tautology but also the probability (probability) of experience induction.

Eastern agriculture in ancient times was more developed and could survive in an agricultural civilization. The interrelation of production and labor needed to establish a harmonious relationship among people. The study of human relations, social relations and moral ethics became the main characteristics of oriental culture, especially of Chinese culture. Eastern culture is also a three-strand confluence: 1. Chinese culture; 2. Indian culture; 3. Japanese culture. Research on the philosophy of humanities in oriental culture is better, the invention of artistic technology is more developed, but research on natural sciences is not developed, Japan paid attention to the study of Western management and natural science and technology after the Meiji Restoration. The methodology of oriental philosophy is inherited, emphasizes inclusiveness and belongs to the Xiang Thinking.

The ancient Greek philosopher Pythagoras believed that everything in the world can be expressed in mathematics simply and directly. Everything has a fixed number, and the number constitutes the beauty and harmony of the world. Pythagoras' philosophy promoted the development of rational thinking. Seeking truth through rational thinking resulted in science. The teaching of Socrates, the ancient Greek philosopher, was to ask students questions that taught them to question existing conclusions, to doubt teachers, and to doubt the correctness of established knowledge, which remained in Western culture. The "doubt everything", proposed by the French Renaissance philosopher Montaigne and the founder of modern Western philosophy Descartes, influenced the way of thinking in modern Western times. The greatest feature of the mind of great scientists, such as Galileo, Copernicus, Newton and Einstein, is the audacity to doubt what is known to be right, and to think of ways to subvert it. In the East, China has Zhouyi, who produced the world's first binary, 64 hexagrams with eight binary numbers to understand everything in the universe, to search for truth in the symbol. The East developed towards understanding thinking, and to find truth by using this method.

The dominant religion in the West is Christianity, and the influential religion in the East is Indian religions (Hinduism, Buddhism, etc.) and Chinese-born Taoism. Christianity is a monotheistic religion and worships the supreme god. However, Hinduism, Buddhism and Taoism are polytheists, and gods are closely linked with natural objects and social activities.

The Renaissance in the West was followed by a revolution, which provided preconditions for the first Industrial Revolution. By the end of the 16th century and the beginning of the 17th century, the British bourgeoisie had matured. Under the influence of the Renaissance movement, people's ideas were liberated and remarkable achievements were made in the field of natural science. Francis Bacon (1561–1626) was both a philosopher and a natural scientist. Bacon believed that the task of philosophy was to go deep into nature, to study and reflect on it, to gain knowledge from it, and to advance science and technology. Bacon put forward the slogan of "Knowledge is power", knowledge is not just empty talk like scholasticism, which is seriously divorced from reality and cannot be replaced by religious belief. The reason why knowledge becomes power is that it can play the role of understanding and utilizing nature, and promote the development of production. Bacon advocated understanding nature through scientific experiments, experiments are observation, recognition and formation of knowledge, the most effective way to obtain new discoveries is induction. Bacon's thought had a positive influence on British natural scientists, which activated the field of British natural science in the 17th century and liberated the productive forces. As a result, Western philosophy began to change from scholasticism to research on natural philosophy, a large number of philosophers became active, and then Western science began to advance by leaps and bounds.

Traditional Confucianism originates from the wisdom of an agricultural civilization. Agricultural civilization gives birth to such cultural characteristics as peace, tolerance, moderation and diversity. Confucianism is the main carrier of this culture. Benevolence, "Be kind and be in good order and harmony" is the core of Confucianism. Benevolence reflected in politics means emphasizing the "rule of virtue". Love is the essence and basic content of benevolence, and this kind of love is consideration of others. Confucianism is inclusive, it is the entry into society and it is progressive; moreover, it has a strong feeling for home and country. Mencius put forward the idea that the people are the most important, the monarch is the least important, and the country is second in importance. The thought of the Confucian Mencius of the people thousands of years ago was put forward earlier than the Western people's idea of equality. The Confucian Zhang Zai: To make a living for the people, to set up a heart for the world, to condition the great knowledge, and to open peace for eternity; The "three Not Afraids" of the Confucian Wang Anshi, in which the "The ancestor's law is insufficient to keep" changes the thought about law; Xu Guangqi, a Confucian scholar of the Ming Dynasty as the Minister of Rituals, studied mathematics, the astronomical calendar and advanced Western ideas from Matteo Ricci, a British missionary, and joined the Catholic Church to become a Catholic; the westernization movement was initiated by a generation of scholars, led by Zeng Guofan. Confucian thought once shone with people-oriented light; however, Confucian thought was reserved and conservative, rather than expanding and innovating. Propaganda is not as good as in ancient times. Whatever a wise man does shall be followed, whatever a wise man does not do is evil. Their words were started with Yao and Shun, and they are complacent and conservative.

After Newton's time, science and technology in the West, especially in England, broke through the religious fetters and became a new thought that dominated the world. This is to use mathematical models to analyze the world and explain the world. However, this way of thinking has indeed made continuous achievements in some fields, especially in the field of natural science. This is why classical mechanics ruled the European scientific community after Newton's time and also penetrated into all parts of the world, including North America and China. Chinese thought breaks through the fetters of Confucianism, which is promoted by the way of Western natural science. Chinese culture (mode of thinking) is an inclusive culture (mode of thinking), if mixed with the fallen things, it will become a conservative culture (mode of thinking); on the contrary, if absorbed with positive things, it will become a progressive culture (mode of thinking). The way of thinking between the East and the West can be complementary. The traditional way of thinking in China pays attention to the whole and to interaction. For example, traditional Chinese medicine is the thinking mode of holism, and pays attention to the balance of yin-yang and internal organs, while the Western way of thinking pays attention to analysis and synthesis, and Western medicine just treats the head when the head aches.

The mode of existence, including mode of production and mode of communication, affects the people's way of thinking. The mode of thinking is just a means, so long as they communicate with each other and practice, they can learn and complement each other. For example, the biggest enemy of innovation is habit. In the course of history, the Jewish nation is full of disasters, lives in no fixed place, is distributed all over the world, must unceasingly adapt to its new environment, solves the new problems, embraces the new methods, and innovates its life style. Today the Jews around the world have achieved remarkable results, and Israel's innovation has been remarkable. After the Meiji Restoration, Japan began to learn science and technology and management methods from the West, and gradually became a modern power. China adheres to the principle of leading the Party and concentrating its efforts on major tasks, adheres to the principle of reform and opening-up, forms the complementary advantages of the lines of thought of the East and the West, and develops better and faster both in economic construction and in science and technology. Many technological innovations, such as quantum communication, have also reached the world's advanced level. In China's primary schools, middle schools and universities, the manner of training of graduate students, besides language courses, other courses are basically in line with the developed countries, communication and production methods are also basically in line with the developed countries, so, the study of the law of things is also in line with the basic way of thinking.

During the course of history, the State of Qi has a rich economy, an advanced political system and advanced military theory (Sun Tzu's Art of War, etc.) so why did it not unify the six countries? On the contrary, the social and military structure of the State of Qin has been transformed into a system capable of fighting together, similar to the united front of the society and the exemplary vanguard role of the army. In the era of cold war weapons, the State of Qin is most effective. Therefore, it is reasonable for the State of Qin to unify the six countries. The key between the Agrarian Revolutionary War and the War of Liberation is to strive for the interests of the peasants in order to form a broad united front. The aim of the Communist Party of China is to serve the people. Therefore, the CPC can unite all forces that can be united to liberate the whole of China. After the founding of New China, through independent wars such as the War to Resist U.S. Aggression and Aid Korea and the development of industries and other production practices, China has united developing countries in Asia, Africa and Latin America (the united front) to jointly develop and it has truly stood erect in the world's national forest, putting forward and practicing the idea that the community of human destiny is a broader united front.

The laws of things are all the same. The complementation of the thinking modes of the East and the West, the traditional wisdom of the West and the methodology of the Marxist outlook on the world, the unique thinking modes and methods in China's excellent traditional culture, the philosophical theories and thinking methods formed by leaders such as Mao Zedong and Deng Xiaoping in their revolution, construction and reform practices, and the methodology of governing the country in Xi Jinping's new era of socialism with Chinese characteristics can all be applied to engineering technology, which has an important methodological enlightenment and reference for the design and construction of highway engineering structures. In years of highway engineering design and construction practice, we pay attention to consciously use philosophical methodology to guide our specific work, and we have achieved great success, and we have accumulated a lot of experience.

# Chapter 2
# Methodology of Highway Engineering

Academician Pan Jiazheng pointed out in his article *Philosophical Thinking in the Construction of Water Conservation* (*Journal of the China Academy of Water Conservation and Hydropower Science*, Volume 1, Issue 1, June 2003): Water conservation engineers have many disciplines to master and cannot spend much energy on philosophical issues. But one's thoughts and deeds are always dominated by one's epistemology and world outlook. If there is any deviation in these aspects, although you have a good heart and master modern scientific and technological knowledge, it will often get half the result with twice the effort and even lead to unexpected consequences. In this way, it is helpful for hydraulic engineers to read some philosophical books. But engineers don't have to read great classics, and somtimes a high-level philosophical essay is not as useful as a proverb. The article vividly summarizes the experience of water conservation projects into eight problems and explains the profound in simple terms, including: (1) the philosophy of looking in the mirror; (2) the philosophy of air travel; (3) the philosophy of taking traditional Chinese medicine; (4) the philosophy of holding eggs; (5) the philosophy of eating arsenic; (6) the philosophy of physical examination; (7) the philosophy of managing children; (8) the philosophy of eating crabs.

Inspired by academician Pan Jiazheng's paper and under the guidance of philosophy professor Feng Jun, the two authors tried to write the methodology of highway engineering structural design and construction by combining the research and practice in that aspect, and re-combed and understood the methodology of highway engineering structural design and construction from a philosophical perspective. We can also learn from the methodology of the *Selected Works of Mao Zedong* (Volume 1–4) and the *Selected Works of Deng Xiaoping* (Volume 1–3) to study the methodology of highway engineering structural design and construction.

**Typical Case 1: Analysis of the "Qiantang River Bridge" Phenomenon**

Mr. Mao Yisheng presided over the construction of the Qiantang River Bridge (Fig. 2.1) with a base period of design of 50 years. According to the design at a

 
H. Zhu and L. Shi, *Methodology of Highway Engineering Structural Design and Construction*, Advanced Topics in Science and Technology in China 59,
https://doi.org/10.1007/978-981-15-6544-1_2

**Fig. 2.1** Photo of the Qiantang River Bridge

speed of 20 km per hour, the original design load railway is Grade E-50 (Hezhong-21.5) and the highway is Grade H-15 (Heqi-11.7). At that time, on average, there were only more than 150 cars and five pairs of trains passing through each day. More than 70 years have passed, and the speed of trains can reach 120 km/h, the speed of cars can reach 100 km/h, and the load of cars can reach 40 tons or even 60 tons. According to today's standards, the Qiantang River Bridge is in a state of extended, over-limited and overloaded service. However, the bridge is still in normal use.

The basic reason why the Qiantang River Bridge can be used normally for more than 70 years is that it can ensure the transmission or transformation of force, deformation and energy according to the design path, avoid the adverse impact on the stable equilibrium of the structure due to the failure to meet the coordination control of deformation, and ensure that the state of the actual use of the structure is basically consistent with the design state, thus ensuring that the bridge is always in a state of stable equilibrium.

**Typical Case 2: Analysis of the "Dujiangyan Irrigation Project" Phenomenon**

Dujiangyan, located on the Minjiang River in the west of the Chengdu Plain, was built at the end of Qin Zhaowang's reign (about 256–251 BC). It is a large-scale water conservation project built by Li Bing and his son, the prefecture of Shu County, on the basis of the excavation of the predecessor Bie Ling. It is made up of parts such as a water diversion fish mouth, Feishayan and Baopingkou. It has been playing a role in flood control and irrigation for more than 2,000 years, and is the largest water conservation project in the world so far, and it is the oldest one, the only one left, still in use and is characterized by dam-free water diversion. It embodies the hard work, courage and wisdom of ancient Chinese working people. The Wenchuan M8 earthquake on May 12, 2008 caused huge losses to Dujiangyan City, but the Dujiangyan Irrigation Project in the severely disaster-stricken area was safe and sound. Aiming at the characteristics and contradictions of the suspended rivers in Minjiang River and Chengdu Plain, it made full use of the local geographical conditions of high northwest and low southeast, and took advantage of the special topography, water pulse and water potential at the river's mouth, guided by favorable conditions, diverted water without a dam, and irrigated by gravity, making dikes, water diversion, flood discharge, sediment discharge and flow control interdependent and integrated into a system. It correctly handled the contradiction between the Xuanjiang River and the Minjiang River and the Chengdu Plain so that they could be unified into a large engineering system and turn water disasters into water conservation, thus ensuring the full play of the comprehensive benefits of flood control, irrigation, water transportation and social water use. The greatest thing about it is that it has lasted for more than 2,250 years, and it provides more and more benefits. Dujiangyan's overall plan is to divide the Minjiang River into two streams, one of which is introduced into the Chengdu Plain, so that both a flood diversion for disaster reduction and for irrigation can be achieved. The main projects include the dike for the diversion of water at the mouth of the river for fishing, the Feishayan spillway and the Baopingkou inlet.

**Philosophy of the Dujiangyan Irrigation Project (Fig. 2.2)**: Dujiangyan has a long history and benefits posterity; in addition to the ingenious engineering layout, the main mystery is to follow the water management guidelines of "taking advantage of the situation and guide, and being appropriate to the time", the management system of "being repaired each year", the principle of river control of "chamfering the dangerous corners and excavating deeply in the straight section", and the Water diversion, sand control, flood discharge management experience and the criterion for weir control of "setting up the fish mouth and repairing the missing, excavating deeply and keeping the weir at a certain height". In more than two thousand years of operation, the Dujiangyan Project has given full play to the potential of the project, and people in the long term practice have accumulated valuable unique experience. The rich cultural connotation reflects the wisdom of the forerunners and the workers, and the formation and development of the Dujianyan Water Culture has fully reflected the correctness and long-term nature of the idea of "Practice is the sole criterion for testing truth". The connotation of the Dujiangyan water culture reflects the whole process of construction, maintenance, management and development, which is one of the important legacies of the development of the human society. This is also an important reason why the United Nations appraises the Dujiangyan Project as an important cultural heritage in the world.

**Typical Case 3: Analysis of the "Zhaozhou Bridge" Phenomenon**

Zhaozhou Bridge (Fig. 2.3), also known as Anji Bridge, was built during the Sui Dynasty (605–618 A.D.) by Li Chun, a famous craftsman. The bridge, with a length

**Fig. 2.2** Photo of the Dujiangyan Irrigation Project

**Fig. 2.3** Photo of the Zhaozhou Bridge

of 64.40 m, a span of 37.02 m, is the world's earliest single-arch open shoulder stone arch bridge with the largest span, which is an important creation in the history of the world's bridge-building.

Zhaozhou Bridge has existed for 1,400 years; it has experienced 10 floods, 8 wars and many earthquakes, especially the Xingtai M7.6 earthquake in 1966. Xingtai is more than 40 km from the bridge, the earthquake there was of a magnitude greater than M4, but Zhaozhou Bridge was not destroyed. Another example is the 1963 flood, which flooded the arch of the dragon mouth, and according to the local elderly people, standing on the bridge you could feel a great tremor, but it was still safe. According to records, Zhaozhou Bridge has been repaired eight times since its completion. Mao Yisheng, a famous bridge expert, said that regardless of the internal structure of the bridge, the mere fact that it has existed for more than 1,300 years shows everything.

In May 1979, a joint investigative group of four units, including the Natural History Group of the Chinese Academy of Sciences, carried out an investigation of the foundation of the Zhaozhou Bridge, which weighs 2,800 t itself; the bridge is based on only a 1.55-meter-high abutment made of five layers of stone bars and built directly on natural sandstone. Such a shallow foundation for a bridge is unbelievable, and Mr. Liang Sicheng, in his investigation in 1933, thought that it was a wall of diamond that could avoid water sweeping through it rather than a foundation to stand the full load of the bridge ring. "To test the circle base, we dug at the foot of the northern circle, but at about 70–80 cm below the river bed, we found the stone wall that had been placed under the circle", he wrote in his report, "The abutment is in a total of five layers, a total of 1.58 m, each layer is slightly longer than the upper layer, and there is no solid foundation below. It is clear that this is only a wall to prevent flow scour, but not the foundation to bear the full load of the bridge ring. Since water can be seen within another 30–40 cm, unless large-scale excavation is carried out, it is impossible to reach the position of the large bridge foundation that we have theorized."

The structure of Zhaozhou Bridge: The main arch is formed by juxtaposing 28 independent arches. When building the bridge, the middle arch was built first, and then the two sides. Each arch is about 35 cm wide and each stone is different in length, ranging from 70 to 109 cm. Each stone is connected by two "waist irons". Each broken arch ring can be repaired separately, because the outer arch rings are more vulnerable to be weathered and damaged, the 5 arch rings at the west side of bridge collapsed during the Ming Dynasty, and they were repaired in the Ming and Qing dynasties (no repair records, but it is obvious that the material of their middle stones is different), the 5 arch rings collapsed at the east side and were repaired in 1955. The 18 arches in the middle were built during the Sui Dynasty.

In order to prevent the arch ring from falling outward, the following methods were also adopted:

1. Narrow at the top and wide at the bottom. In order to prevent the arch ring from tilting outward, the foot of the arch was built 9.5 m wide and the top of the arch was built 9 m wide.
2. Twenty-eight arch rings were transversely connected with 9 iron connecting rods. A few of iron connecting rods were used with holes at one end and a hook at the other end to connect the 28 arch rings within the arch ring. At the outermost arch ring, there is an iron head (as the picture below). There are five pull bars on the main arch, and there is one pull bar on each small arch.
3. Each block of stone was reinforced with sawn-like waist iron, and two waist rails were used at the junction of each piece of stone.
4. Six stone hooks were used to pull the outer bridge arch. A 1.8 m long and 5 cm long curved stone was used on the head to place it on the outer side of the bridge

arch. The extra part was used to pull the bridge arch inward. Because each bridge arch is 35 cm wide, the stone can hold down the 5 outside bridge arches. However, its function is limited. For example, the arches on the east and west sides of the bridge all collapsed, and the five arches on the outside of the two sides were all repaired afterwards. When the bridge was repaired in 1956, the arch crown was reinforced with steel mesh.

**Typical Case 4: The Ranking of Difficulty in the Structural Analysis of Highway Engineering**

Generally, the degree of difficulty of highway engineering structure analysis is ranked according to the tunnel, bridge, slope, road surface and so on, but from the point of view of comprehensive research and analysis, there are interaction relations between the tunnel structure and the surrounding rock. A lot of experiments and theoretical research have been done regarding the surrounding rock and the lining. Even if some parameters are wrong, that has little effect on the safety of the structure of the tunnel. The bridge structure's external load is relatively clear, the bridge structure's analysis is relatively mature, the value of the parameter is also relatively stable, and the degree of accuracy of the analysis of the bridge structure is high. The interaction between the rock and the soil of the structure of the slope is much weaker than that of the tunnel. Although a lot of experiments and theoretical studies have been carried out, the analysis of the structure of the slope is still far from comprehensive research. Practice has proved that there have been many collapses after the construction and operation of the structure of the slope. The research on the structure of the pavement is focused on material and technology. At present, there is not enough research on the interaction between subgrade and pavement and the law of the stress and deformation of the structure of the road. Practice has proved that there are many cracks and much breakage after the construction of the structure of the pavement, but the risk to driving safety is relatively small. Based on the method of comprehensive analysis, it is found that in the ranking of the difficulty of tunnel, bridge, slope and pavement, pavement is the most difficult, and then slope, bridge and tunnel, which are relatively simple. Practice has proved that after the construction and operation of the structures of the pavement, slope, bridge and tunnel, there are cracks and breakage. This is contrary to the common view that tunnels are the most difficult, followed by bridges and side slopes, and the relatively simple order of road surfaces.

When Mr. Mao Yisheng built the Qiantang River Bridge, he mainly used the combination line of thought and structural mechanics characterized by rational analysis. The Dujiangyan Irrigation Project built by Li Bing and his son and Li Chun and the Zhaozhou Bridge era were mainly an epiphany or a holistic line of thought characterized by rational intuition. Mr. Mao Yisheng studied in the West and his way of thinking is both Eastern and Western. Li Bing and his son and Li Chun have mainly oriental manners of thinking. Although their thinking methods (means) are different or complementary, they have all conducted a comprehensive study on the fundamental problem of the laws of stress and deformation of different engineering structures, and have finally achieved their goals. That is, the manner of thinking is a means and it can change, and can be complemented by mutual learning and reference. That is, in actual engineering practice, as long as the correct mode of thinking is grasped, the fundamental problem of the law of stress and deformation of the engineering structure can be better solved.

Some substances (sand, water, silt, etc.) are integrated and require external control conditions; the design and construction of highway engineering should first study comprehensively to avoid disasters and common forces, and then analyze and study the forces on favorable composite structures. For example, the Sichuan–Tibet Highway is a modern upgraded version of the ancient Sichuan–Tibet line known as the "ancient tea-horse road". In early 1950, Chairman Mao Zedong instructed troops entering Tibet to "build roads while marching" to liberate Tibet. In more than 4 years, the Sichuan–Tibet Highway passed through 14 mountain ranges including the Erlang Mountain, the Zheduo Mountain, the Que'er Mountain and the Shergyla Mountain, it crossed the Minjiang River, the Dadu River, the Jinsha River, the Nujiang River, the Lhasa River and many other rivers, and it crossed eight major fault zones, including Longmen Mountain, the Qingni Cave, the Lancang River and Tongmai. The project is unprecedented in the history of road construction in the world, overcoming various difficulties. When the Sichuan–Tibet Highway was built, it was restricted by many factors such as historical conditions and economic and technical level. The highway was built in a short period of time, with a low degree of engineering and rough construction. It was basically an emergency military highway. In addition, the hydro-meteorological and topographical and geological conditions along the highway were very complex, and various mountain disasters occurred frequently. Therefore, car blocking and cut-offs occurred frequently. Since 1985, the state began to repair the Sichuan–Tibet Highway, but basically along the old road to expand it into a three-lane highway. From 2012 to April 2016, the "intestinal obstruction" project of Tongmai was rebuilt, including the newly-built Tongmai extra-large bridge, the Pailonggou extra-large bridge and several tunnels. The original dangerous section of more than 20 km was reduced to over 5 km, and replaced by a tunnel under a mountain and a bridge across a river, namely, and the renovation and reconstruction project of the Tongmai section of the Sichuan–Tibet Highway with "five tunnels and two bridges" as the main part was officially opened to traffic. The world-famous blocked section of Tongmai in Tibet along the Sichuan–Tibet Highway became a historical and a natural barrier became a thoroughfare. Previously, crossing the Sichuan–Tibet Highway was called "the soul is in heaven and the body is in hell." It is now called "the body walks in heaven, the mind purifies in it". Since 2002, the author has participated in the reconstruction of the Sichuan–Tibet Highway, collected a large amount of data and photos, and carried out the research work on highway engineering structures. The Lhasa-Nyingchi highway in Fig. 2.4, where the route and structural types to avoid the impact of disaster due to water damage are comprehensively studied, and then the structure of the design and construction of the highway is specifically analyzed. However, the Lhasa-Nyingchi tertiary highway is limited to the influence of capital and equipment, and cannot avoid the water damage to its subgrade. The Qinghai–Tibet railway subgrade shown in Fig. 2.5 has measures to resist freezing and thawing. Bridges are substituted for roads in special sections to eliminate freezing

**Fig. 2.4** The Lhasa-Nyingchi Expressway and Tertiary Highways have different abilities to cope with water damage, and the stability of highway conditions is also different

**Fig. 2.4** (continued)

and thawing effects and ensure the operation of the railway all year round. However, the Qinghai–Tibet highway subgrade has insufficient measures to resist freezing and thawing, resulting in much damage to the pavement.

Figure 2.4 shows that the Lhasa-Nyingchi Expressway replaces the road with a bridge or a higher subgrade to avoid the riverside, thus avoiding the phenomena of water damage, frost boiling, freezing and thawing, etc. On the other hand, the Sichuan–Tibet Highway is a tertiary highway along the river, with a low subgrade and is vulnerable to disasters such as water damage, frost boiling, freezing and thawing.

Figure 2.5 shows that the Qinghai–Tibet Railway is very close to the Qinghai–Tibet Highway. The Qinghai–Tibet Railway replaces the road with a bridge or the subgrade is relatively high, and the bottom of the railway is provided with a layer of broken stone cushion for ventilation and temperature equalization, effectively resisting freeze-thaw, frost boiling and other phenomena. However, the Qinghai–Tibet highway has a low subgrade and no broken stone cushion at the bottom, so it is prone to freeze-thaw, frost boiling and other disasters.

Von Karman, a famous mechanic, said that "science is to discover what already exists, while engineering is to create what has never existed in the world". The law is inherent and can only be discovered through observation, experimentation and statistics, but engineering construction must conform to the basic law. There is an implicit connection (easy to ignore) between calculus, engineering mechanics-element combination $\sum$ or $\int$ differentiation that keeps the original whole, namely deformation coordination control. However, some engineering structures do not fully meet the deformation coordination theory and need control methods. There is an

**Fig. 2.5** The Qinghai–Tibet Railway and Tertiary Highways have different abilities to deal with freezing and thawing, and the stability of the state of the subgrade is also different

Fig. 2.5 (continued)

engineering structure question about engineering mechanics. There is an effective connection between structures, which must satisfy the coordinated control of deformation; In particular, for complex problems of the engineering structure, there is a need to study not only external influences and ontological problems, but also internal contradictions and problems of connection. The deformation coordination hypothesis of engineering structures m and s may be satisfied, some engineering structures may not satisfy the deformation coordination assumption of m and s. Therefore, in the design of engineering structures, the effective connection between structures must maintain the coordinated control of deformation.

When the organizational structure of the engineering structure is reasonable, i.e. the deformation coordination control is satisfied, the state of stress and deformation of the unit $A_i$,... and organizational node $x_i$,... is coordinated. The material properties and microstructure of the engineering structure are unchanged during the process of stress, which accords with the conditions of applying calculus and engineering mechanics. Engineering mechanics can be applied to solve the stress and deformation problem of the engineering structure.

When the organizational structure of the engineering structure is unreasonable or has a gap, namely it does not satisfy the deformation coordination control, the state of stress and deformation of the unit $A_{i,...}$ and organizational node $x_i$,... is incongruous. The material properties and micro-structure of the engineering structure will change during the process of stress, and the change mode and law are unknown, which does not meet the conditions of applying calculus and engineering mechanics. If engineering mechanics is applied to solve the problems of engineering structural mechanics and deformation, there will be deviations or even errors. The design and construction of the engineering structure should adopt the method of overall control and the grasping of details, that is, when the engineering structure is in a state of metastable equilibrium, the reasonable structure should be designed or auxiliary measures should be adopted through the specification, analogy and test determination of the measures of the engineering structural construction to control the stability of the state of stress and deformation of the engineering structure, that is, after the "metastable structure" is converted into the "stable structure", the engineering mechanics should be applied to solve the problems of the mechanics and deformation of the engineering structure. For example, in Fig. 2.6, the measures for the protection of the road are not in place or are wrong, resulting in insufficient stability of the subgrade condition and making it easy to cause traffic accidents. In Fig. 2.7, the highway is protected by an open tunnel or a shed tunnel to avoid the hazards of collapse or sand slide slope. That is to say, only by studying the regularity of things can we correctly grasp the methods for the management of disasters of highways and then correctly design and construct highway structures.

Figure 2.6 shows that the level of the reconstruction of the Sichuan–Tibet Highway is low or not in place, and it is easy for disasters such as water damage and the like to occur along the riverside, and it is also easy for disasters such as collapsing and ice jams and the like to occur along the mountain side, as well as disasters such as freeze-thaw, frost boiling and the like to occur at the lower part of the subgrade at the high altitude, and even accidental damage and the like may easily occur due to unreasonable design.

Figure 2.7 shows that the Sichuan–Tibet Highway has been rebuilt to improve the standard, tunnels and open cut tunnels have been adopted to avoid many unexpected disasters so as to ensure the smooth flow of the highway.

For example, the area severely damaged by the Wenchuan M8 earthquake on May 12, 2008 exceeded 100,000 km$^2$, which was the most destructive earthquake since the founding of new China (1949) and the most deadly earthquake after the Tangshan

**Fig. 2.6** Insufficient stability of the highway subgrade condition makes it easy to cause traffic accidents

**Fig. 2.6** (continued)

Fig. 2.6 (continued)

**Fig. 2.6** (continued)

**Fig. 2.6** (continued)

earthquake (July 28, 1976). Among them, the loss of houses was very great, with the loss of houses, and the houses of urban residents accounted for 27.4% of the total loss. The loss of roads, bridges and other urban infrastructures accounted for 21.9% of the total loss. On May 16, the author took part in the traffic rescue work in Guangyuan City with Qingchuan County as the center, and also took part in the road reconstruction work in Qingchuan County later, collected a large amount of data and photos and carried out in-depth research work. It was found that railways and expressways were transporting disaster relief materials, while many national, provincial and rural roads with relatively low grades needed to be cleared and dredged, and even steel bridges for war preparedness were set up. Villages and towns with houses of a wallboard structure were basically completely collapsed over and suffered casualties, while houses with a frame structure were damaged or partially cracked, but there were no casualties. Even huge stones falling from mountains that impacted the one-storey high houses with a frame structure only damaged the walls. After the rescue work was completed, the author went to the Dujiangyan Irrigation Project to investigate the earthquake damage. Except for partial damage repair, it was in normal operations. Later, the authors went to Japan and Chile to inspect the earthquake work. They also summarized the experience and lessons of the previous earthquake and gradually improved the engineering structure to achieve better seismic performance. For example, staff members said that when the Chilean earthquake occurred (M8), people on the street ran into houses to avoid being hurt by falling objects. However, when the earthquake occurred in our country, the people in the houses were asked to evacuate to the open area. The difference is the ability of bridges, houses and other

**Fig. 2.7** Highways shall be protected by an open cut tunnel or a shed tunnel to avoid the hazards of collapse or sand slide slope

**Fig. 2.7** (continued)

**Fig. 2.7** (continued)

engineering structures to resist earthquakes. Examples are as follows (Figs. 2.8, 2.9 and 2.10).

The important part is the rationality of the construction of the engineering structure and the adaptability of the geological environment, which causes the structure to always be in a state of "stable equilibrium and deformation coordination control", namely the state of force safety.

For example, through an on-the-spot investigation of highway tunnels in the Wenchuan earthquake area and image analysis of tunnels from media, some common aspects and laws were found and preliminary suggestions on seismic design and reinforcement methods of tunnels in the earthquake area were put forward for the reference of designers, construction personnel and decision makers.

In the discussion on the selection of the site, an old gentleman noticed that the tunnel damage is less when it is built in a mountain valley (when the mountain terrain is symmetrical), and more when it is built in the middle of a slope (when the mountain terrain is biased). At present, science and technology are relatively developed, and the cost of tunneling is not necessarily higher than that of constructing expressways, which will have an impact on the future plan layout and the planning of the urban and rural system.

(a) The destructive situation of the unreasonable structure of the presidential palace in Haiti

(b) The situation of minor damage to housing with a reasonable kind of structure in Chile

**Fig. 2.8** Comparison of damage to buildings in Haiti and Chile in the M8 Earthquake

(a) The houses with a frame structure were basically left intact after being hit by boulders during the Wenchuan Earthquake

(b) Houses with a wall slab structure were completely destroyed in the Wenchuan Earthquake

**Fig. 2.9** Comparison of building damage in the Wenchuan Earthquake

(a) The lining of the tunnel entrance (the mountain terrain is symmetrical) is basically intact.

(b) The situation of the lining of the tunnel exit being cracked (the mountain terrain is biased)

**Fig. 2.10** Comparison of the lining of tunnel entrance (the mountain terrain is symmetrical) and exit (the mountain terrain is biased) after the earthquake

**Fig. 2.11** Entrance of the Jianmen Pass Tunnel in Jiange County (the mountain terrain is symmetrical)

I. When the mountain terrain is symmetrical

Under the condition that the entrance and exit of the tunnel pass through the mountain with symmetrical terrain (Figs. 2.11 and 2.12), the process of seismic wave propagation and the tunnel stress is relatively simple and uniform, and the lining is relatively intact.

II. Slope environment

In the slope environment, on the one hand, the process of the propagation of seismic waves will become more complicated, resulting in the concentration of local stress in the slope body, on the other hand, short-term local tensile stress will occur in the slope body due to the temporary air conditions of the slope. This kind of local stress concentration and tensile stress will cause damage or even destruction to underground engineering structures such as tunnels.

In Fig. 2.13, the bias of the mountain terrain of Jinzishan Tunnel is relatively small, and the impact of the earthquake is relatively light. The transverse cracks are reinforced with plum blossom-shaped anchor bolts, and the cracks are sealed with epoxy resin.

In Fig. 2.14, the mountain body of the Jiujiaya Tunnel has a large topographic bias, and the mountain body is broken and has a great impact from the earthquake. The

**Fig. 2.12** Entrance of the Jiujiaya Tunnel in Qingchuan County (the mountain terrain is symmetrical)

inlet linings of the tunnel should be transformed into a reinforced concrete structure piece by piece.

In Fig. 2.15, the mountain terrain of Sanpanzi Tunnel is relatively biased, but the mountain is relatively complete and the earthquake impact is moderate. The part with a transverse crack should be reinforced with a plum blossom-shaped anchor bolt and hung with steel mesh to be sprayed with 5 cm of concrete.

In Fig. 2.16, the mountain terrain at the entrance to the Caopo Tunnel is greatly biased, and the broken part of the mountain basically collapses due to an earthquake, which affects the consequent relationship of the bridge-tunnel structure at the entrance to the tunnel.

In Fig. 2.17, the cave entrance of the Baoji–Chengdu Railway No.109 Tunnel collapsed due to the earthquake, which caused the oil tank train to burn up completely and cause an explosion, resulting in complete destruction of the tunnel lining and collapse of the local surrounding rock. The treatment methods for the maintenance and reinforcement are as follows: (1) The rescue plan will be carried out in parallel with the restoration of existing lines and the change of lines. The plan for the protection of the existing line reinforcement is conducive to the rapid opening of traffic to meet the needs of the transportation of disaster relief materials. After the existing line was hit by the disaster, the recovery plan can only be temporary; in order to ensure long-term operations, the line must be rebuilt. The rerouting plan adopts a

(a)

(b)

**Fig. 2.13** Entrance (**a**) and exit (**b**) of the Jinzishan Tunnel in Jiange County (the mountain terrain is biased)

(a)

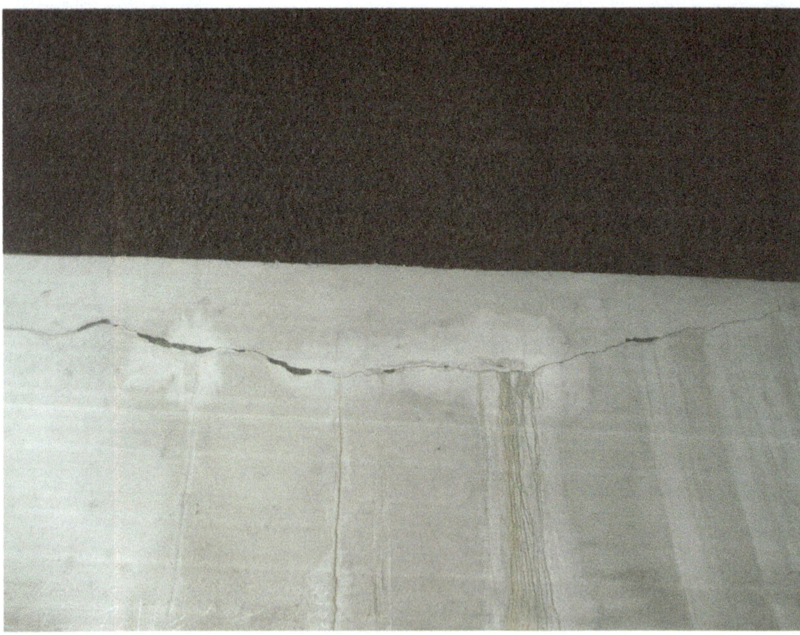

(b)

**Fig. 2.14** Jiujiaya Tunnel in Qingchuan County (the mountain terrain is biased). **a** Entrance; **b** and **c** damage or even destruction of the structure of the entrance

(c)

**Fig. 2.14** (continued)

short tunnel, which is located opposite the Jialing River of tunnel 109. The rerouting is about 2.08 km long and consists of one tunnel and two bridges, across the Jialing River twice. Among them, the tunnel is 860 m, the length of the first bridge across the river is 248 m, while the second is 140 m. (2) The tunnel will be strengthened by a shotcrete anchor in the sections that were slightly damaged by the earthquake and fire, and the serious sections will be treated by comprehensive measures such as the erection of a steel arch and shotcrete anchor reinforcement. (3) Strengthening of the collapsed mountain outside the tunnel, cleaning up damaged car bodies inside the tunnel, hanging nets and shotcreting reinforcement and other work were carried out in a three-dimensional manner and were comprehensively promoted. At present, the work of repairing the deformed tunnel, reinforcing the broken roof beam, and rushing to build the anti-collapse shed tunnels has been completed in an all-round way.

III. Special problems

In Fig. 2.18, the exit of the newly-built Jianmen Pass Tunnel is relatively thin in the area of the collapse of the mountain and has been greatly impacted by the earthquake. Its lining should be transformed into a reinforced concrete structure piece by piece.

a

b

Fig. 2.15 Entrance (a) and exit (b) of the Sanpanzi Tunnel in Qingchuan County (the mountain terrain is biased)

**Fig. 2.16** Entrance of the G317 Caopo Tunnel (the mountain terrain is biased)

In Fig. 2.19, the mountain in the middle of the newly-built Jianmen Pass Tunnel has breakage and seepage. Due to the earthquake, the mountain in the shallow section is cracked. The design of the tunnel was modified to dig out the mountain in the shallow section, symmetrically backfill the whole with clay, and pave the surface with masonry to solve the problem of water leakage from the mountain in the shallow section.

The enlightenment of the three magic weapons of the Chinese revolution in engineering application is as follows: The united front corresponds to the common stress on all parts of the structure, the armed struggle corresponds to the equilibrium and stable stress of the structure, and the party building corresponds to the stress of the main structure and the subsidiary parts of the organization. Therefore, it is very important for the engineering structure to bear the common force and the main force and organization of the main structure. For example, in the face of construction difficulties such as collapse, water leakage, excessive formation deformation and the like in underground engineering construction caused by crossing complex engineering environment and sections with unfavorable geological conditions, different structural measures are adopted according to the deformation coordination control method and different states of rock and soil masses (surrounding rocks) to effectively control effective bearing structural layers formed in advance or immediately or inherently, so that the rock and soil masses and support structures can form a stable combined system against the deformation. Moreover, every part of the rock and soil mass (surrounding rock) plays a role in balancing the underground structural system, transforming the load or burden borne by the structure into resistance or resources

**Fig. 2.17** Rescue and reinforcement of the Baoji–Chengdu Railway 109 Tunnel

Fig. 2.17 (continued)

**Fig. 2.18** Exit of the newly-built Jiangmen Pass Tunnel in Jiange County (landslide)

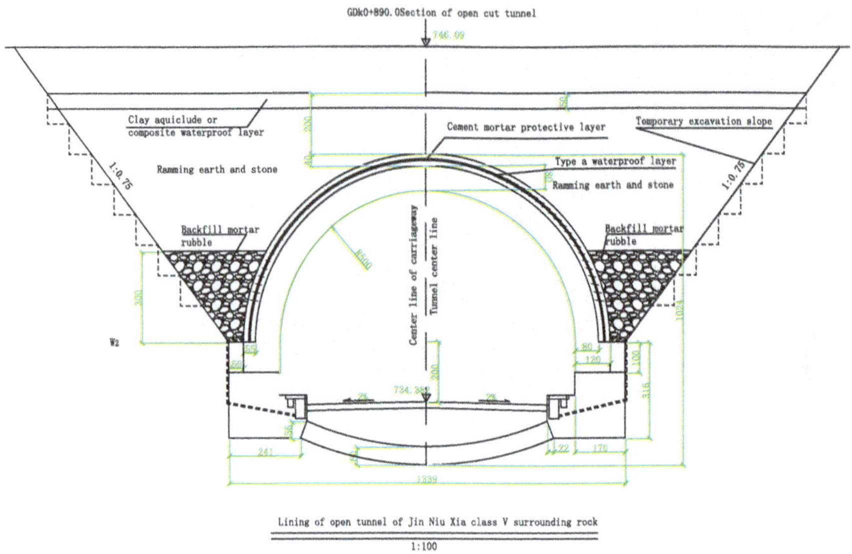


**Fig. 2.19** Middle part of the newly-built Jianmen Pass Tunnel in Jiange County (mountain breakage and water seepage)

that play a role in structural equilibrium, so as to reasonably play the self-bearing capacity of the rock and soil mass. In this way, the joint stress of the structure + rock and soil mass (surrounding rock) is in equilibrium and stable (corresponding to the united front concept), in which the structure or rock and soil mass plays the main and organizational roles (corresponding to the concept of the Party's construction), effectively exerts the joint stress of the surrounding rock, converts the load or burden borne by the structure into the resistance or resources that play the role of structural equilibrium, and achieves the reasonable self-bearing capacity of the rock and soil mass under the New Austrian Tunneling Method.

Any part of an engineering structure should be regarded as a resource rather than a burden. On the basis of existing mechanics, a stable equilibrium can be achieved by giving full play to the role of any part by means of a reasonable structure. The better way to achieve this is the coordinated control method of structural deformation.

The components of ice, snow, water, water vapor are all $H_2O$, but they have different structures and a different stable equilibrium condition of the forces. Among them, ice is approximately in a stable equilibrium, snow is in an approximately metastable equilibrium, while water and water vapor are in an approximately unstable equilibrium. If snow, water and steam are controlled in a stable state by a container, they can also be in a stable equilibrium state. Naturally, they have different methods of analysis in different states. This method of analysis is not only based on mechanical methods but it also springs from mechanical thinking. The philosophy, mechanics, control and other comprehensive methods of analysis for thinking about the analysis

**Fig. 2.20** Photo of block collapse risk at the slope of the entrance to the tunnel

of engineering structures are similar to those based on and springing from industries to better plan the development of industries. (1.1) For engineering structures that involve interaction with strata, such as tunnels, foundation pits, side slopes and soft soil subgrade, it is the most economical solution to adopt effective engineering measures to maintain the original state of stress and deformation of the strata. (1.2) For any engineering structural system that has gaps in satisfying reasonable conditions of structural or deformation coordination control, for example, the engineering structural state is in a state of a metastable equilibrium, and the method of overall control and detailed control is the most economical method for engineering structural design and construction. The above two kinds of situations are to design reasonable structures or adopt auxiliary measures through specifications, analogy and test determination of engineering structural construction measures, i.e. adopt the method of deformation coordination control to control the stability of the state of mechanical deformation of the engineering structure; (2.1) For engineering structures that involve the formation of water damage, such as engineering water damage caused by bridge foundation erosion and gully erosion in mountain rivers, effective engineering measures to control water flow, velocity and erosion time are the most economical solutions. That is, referring to the concept of the impulse theorem, the key factors for macro control of water damage can be found, and the problem of erosion or water damage can be solved by constructing effective engineering measures. For example, many mountain highway tunnels built in China during the 1980s and 1990s adopted the method of shortening the length of the tunnel and forming a cut slope at the entrance so as to save money. But the phenomenon of the risk of the tunnel entrance slope sliding was also formed as shown in Fig. 2.20; after 20 years of operations, we adopted the plan of extending the open cut tunnel as shown in Fig. 2.21 to meet the common stress requirement and prevent the collapse of the mountain.

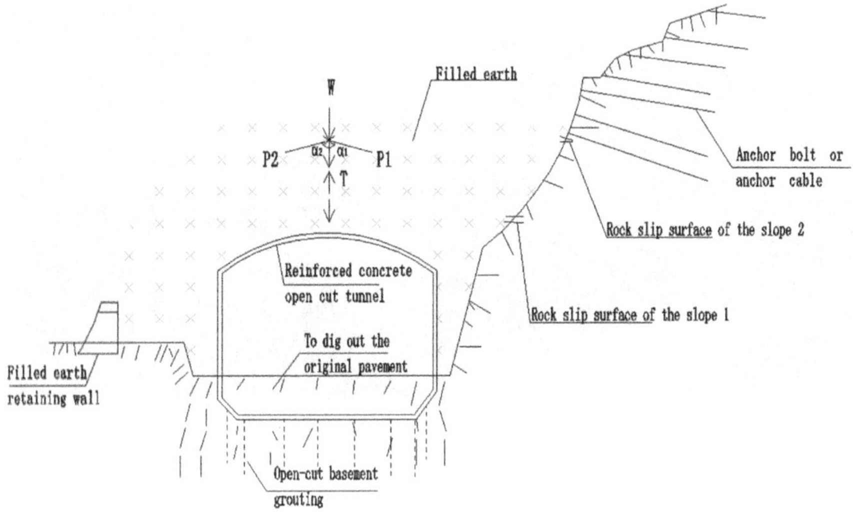


**Fig. 2.21** Schematic diagram of the extension of the open cut tunnel

Why is it that many solutions for the "bump at bridgehead" of a soft clay slag subgrade only have some improvement and have little curative effect? These two key points have not been resolved:

I. Characteristics of Soft Soil Foundation: (This is also the two major reasons that the dynamic load of traffic easily causes settlement of the soft soil slag subgrade after construction!)

   (1) Thixotropy: Once soft soil is disturbed, the structure is destroyed, and its strength rapidly decreases or becomes diluted quickly. Therefore, the soft soil slag foundation under a vibration load is prone to side sliding, settlement and extrusion on both sides of the bottom surface.
   (2) Rheology: This refers to the characteristic that the deformation of soil increases with time under the constant action of a certain load. Long-term strength is much smaller than the instantaneous strength. This is very unfavorable to the stability of the subgrade, slope, embankment, wharf and so on.

II. Code for the Design of the Subgrade: (Why is long-term compression settlement difficult to control after the construction of slag subgrade in soft soil?)

   (1) The main consolidation is the process of compression (compression settlement during construction) caused by the gradual discharge of pore water after compression under the condition of lateral limit.

(2) Secondary consolidation is a process in which the volume of soil decreases with time after the completion of primary consolidation after compression under lateral conditions. It can also be understood as the rheological deformation of soil (long-term compression settlement after construction).
(3) Total settlement $S = mS_c$ (m is the sedimentation coefficient, 1.1–1.7; $S_c$ is the main consolidation settlement). Due to the rheological effect of soft soil, the secondary consolidation settlement is stable after a long time and its value is much larger than that of the sedimentation coefficient. This is why it is difficult to control the long-term compression settlement after construction of soft soil slag subgrade.

In order to solve the above problems, the author's team adopted the lower threshold of soft clay rheology to control the post-construction settlement of soft soil slag subgrade for the first time at home and abroad, and improved the designing method of soft clay slag subgrade according to the coordinated control method of structural deformation first established by the team at home and abroad: A. If the lower threshold of soft clay rheology is greater than the stress value at the bottom of the soft clay slag subgrade, the post-construction settlement of the soft clay slag subgrade can be controlled; B. If the lower threshold of soft soil rheology is less than the stress value at the bottom of the soft soil slag subgrade, engineering measures, such as light subgrade, adding a lower partition plate or frame lattice to the slag subgrade and pile transition combination technology, should be taken to control the post-construction settlement of the soft soil slag subgrade within the allowable range and achieve the purpose of solving the bump phenomenon at the bridgehead.

Through more than ten years of efforts, the author's team has successfully solved the problem of "bump at bridgehead" of nine roads (one in Beilun, Ningbo, and one in Zhenhai, Ningbo; seven in Jiaxing), and has been authorized by many countries invention patents, published a number of books at home and abroad. The Zhejiang Provincial Association of Technology Brokers with academician Sun Jun as the head of the appraisal committee decided that the results of the research reached the international advanced level, China's traffic newspaper and traffic tourism guide reported it in their headline editions.

For example, as shown in Figs. 2.22 and 2.23, in view of the two problems existing in the methods for soft soil subgrade analysis and treatment, the traditional soft soil subgrade analysis theory adopts deterministic methods [springs, clay pots, sliders, etc.] to simulate the characteristics of the uncertain movement of soft soil subgrade [particles], and the international community generally believes that the calculation of the force of the soft soil subgrade is accurate and the calculation error of deformation is large, which is inconsistent with the functional relationship $y = f(x)$. Although the theoretical logic is rigorous, it does not conform to dialectical thinking. Only by controlling the regular movement of soft soil particles technically, that is, simply controlling the integrity of the subgrade, can the results of the deformation calculation of the soft soil subgrade be within the allowable range of engineering. On the basis of summing up experience and dialectical thinking, the design of the soft soil subgrade is improved with the aim of controlling the integrity of the state of stress

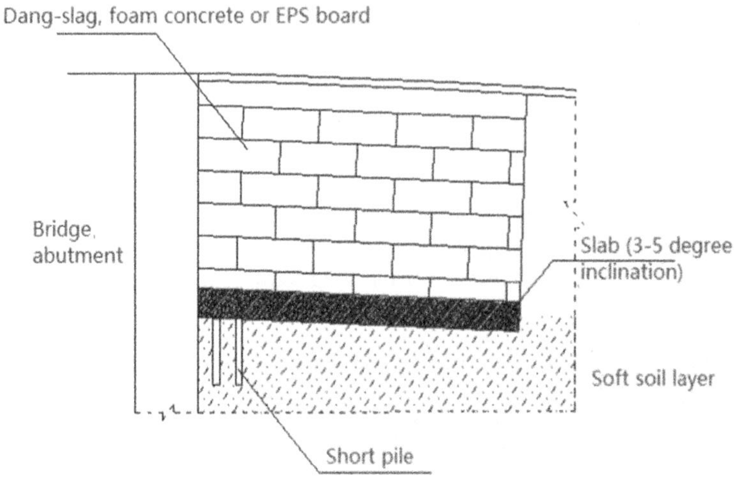


**Fig. 2.22** Bottom plate or frame and short piles to control the post-construction settlement of the subgrade

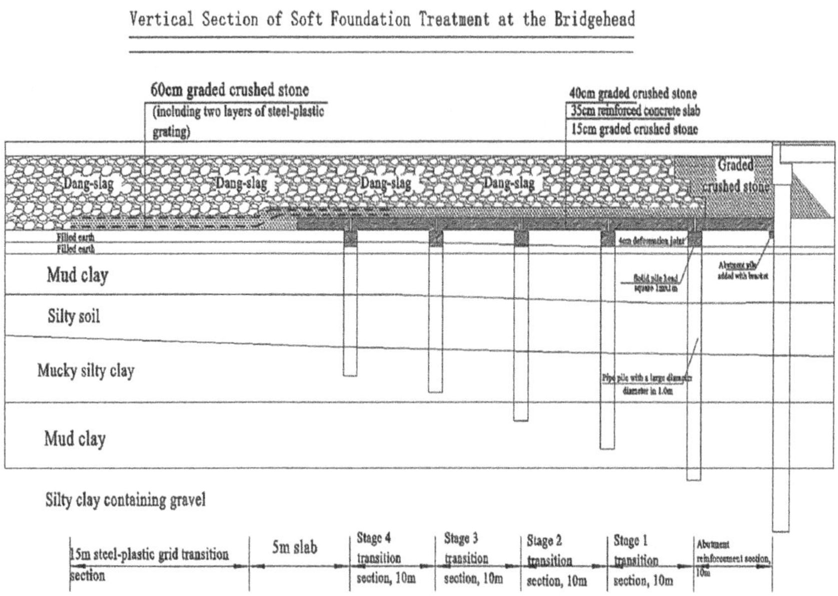


**Fig. 2.23** Bottom plate or frame and longitudinal transitional pile foundation to control the post-construction settlement of the subgrade

and deformation of the soft soil subgrade and the unstable continuous settlement caused by soft soil rheology. According to the coordinated control method of structural deformation, the designing method of the soft soil subgrade has been improved. In the design, the integrity and longitudinal transition of the state of stress and deformation of the subgrade are controlled. The key technology is adopted to control the bottom plate or frame and the foundation of the longitudinal transition pile of subgrade settlement after construction. Among them, the length of the longitudinal transition of the pile is related to the gradient height of the bridgehead subgrade. For light materials, attention should be paid to the control of layered integrity and the structural layer with a uniform distribution of vehicle load. In this way, the post-construction settlement of the bridge-head subgrade can be controlled within the allowable range of the project, and the purpose of treating the "bump at bridgehead" of highway soft soil subgrade of can be achieved.

# Chapter 3
# Comprehensive Research and Analysis

To learn from Mao Zedong's thoughts on the *On Protracted War*, we should first make a comprehensive analysis of the situation of the world and the relationship between China and Japan, and answer the question: Will China die? No, the final victory is China's; Can China win quickly? You can't win fast. The Second Sino-Japanese War is the long game. What should we do? The first was the establishment of the Chinese Anti-Japanese United Front; the second was the establishment of the International Anti-Japanese United Front; and the third was the rise of the Revolutionary Movement of the Japanese people and the people in their colonies. As far as the Second Sino-Japanese War is concerned, of the three conditions, the great unity of the Chinese people is subjective and major. Mao Zedong used a synthetic approach, which solved the strategic problems of the Second Sino-Japanese War, strengthened national confidence in winning this war, and pointed out the way to specific campaigns. This method can also be used in engineering mechanics and highway engineering structural design and construction.

Engineering mechanics is based on the determination of material properties and microstructure, but some engineering structures will change in the process of loading, and the mode and rule of that change are unknown. For example: unfavorable structures or composite engineering structures such as flexible bending members, broken surrounding rocks, soft soil foundation, etc.; therefore, the application of engineering mechanics to solve engineering structural problems is conditional. That is to say, comprehensive research should be carried out first. When the engineering structure does not meet the above conditions, engineering measures should be taken to control the stress and the state of deformation of the engineering structure, and then an analysis of the engineering mechanics should be carried out. For example, Karl Terzaghi, the founder of soil mechanics, concluded in his later years that "soil mechanics is not so much science as technology. The most important purpose is to reasonably complete the design and construction of geotechnical engineering, and mathematics and mechanics are just the means"; for another example, Western economic theory cannot analyze the history of Western industrialization, nor the achievements of



H. Zhu and L. Shi, *Methodology of Highway Engineering Structural Design and Construction*, Advanced Topics in Science and Technology in China 59,
https://doi.org/10.1007/978-981-15-6544-1_3

China's reform and opening-up, and 90% of the content of economic theory is more difficult for enterprises to apply, just limited to the mathematical tools of many economists, without grounding in theory. Therefore, the comprehensive methods of research and analysis are inseparable. To solve the structural problems of highway engineering, it is necessary to carry out mathematical and mechanical analysis on the basis of comprehensive research, so as to better grasp the fundamental problem of the law of stress and deformation of the engineering structure and provide a guarantee for the correct solution of the structural problems of highway engineering.

The highway engineering structure serves the needs of national economic and social development, and must be able to withstand the harsh natural environment, complex geological and hydrological conditions and cyclic loads in the process of passing through mountains and rivers. The design and construction of the linear engineering structure must be based on the comprehensive study of the function of the engineering structure, it must grasp the relationship between the whole and the part, and solve the practical problems:

(1) Comprehensive research (strategic problem): familiar with the environment and the load, and give full play to the supporting capacity of composite structures

① Determine the situation (determine the motion state and trend of the object, and transform the factors of uncertainty into problems of certainty).

② Control hazards (principle of minimum energy consumption, rationality of force medium, coordinated control of deformation, adaptation to nature and transformation of nature—archaeological discovery—the climate of the Hexi Corridor changed from humid to arid about 3,700 years ago).

For example: In the Pailong section of the Sichuan–Tibet Highway (Fig. 3.1), there is a metastable Tianchi on the mountain, and sometimes there is a flow of water impacting the safety of the road, such as the Mozhugongka landslide and 102 landslide; disasters such as narrow roads, sharp bends, dangerous rocks, rolling stones, collapses, landslides and mudslides at the foot of the mountain can occur, which are the most serious "cecum" sections blocking the traffic, and there is a lack of comprehensive research in history and the reconstruction of the original roads aimed at keeping the traffic flowing did not fundamentally solve the problem at all. In 2012, through comprehensive research, four tunnels in Pailong and the Parlung and Tongmai super large bridges were rebuilt to avoid the mountain disasters such as narrow roads, sharp bends, dangerous rocks, rolling stones, collapses, landslides, mudslides, etc., which made the former sections with frequent natural disasters and accidents become a smooth road, as shown in Figs. 3.1a, b. This is a case study to solve the complex problems of highway engineering structures.

Figure 3.1a compared the ability of bridges and tunnels to deal with disasters such as water damage before and after the reconstruction of the Pailong section. Figure 3.1b shows that before the reconstruction of the Sichuan–Tibet Highway, due to a lack of funds, the design level was low and not in place, which led to disasters such as water damage, collapses, etc., and dangerous

**Fig. 3.1** **a** Pictures of Tianchi in the presence of some metastable states in the mountains taken from an aircraft. **b** Photos of some dangerous road sections before the reconstruction of the Sichuan–Tibet Highway. **c** Road condition photos of some sections with smooth road for traffic after the reconstruction of the Sichuan–Tibet Highway. **d** Comparison of road conditions and disaster response capacity of several tunnels or shed tunnels before and after the reconstruction

**Fig. 3.1** (continued)

Fig. 3.1 (continued)

(b)

Reconstruction of 102 landslide group and Tongmai to 105 road section on the Sichuan-Tibet Highway (Tibet border) of China National Highway 318

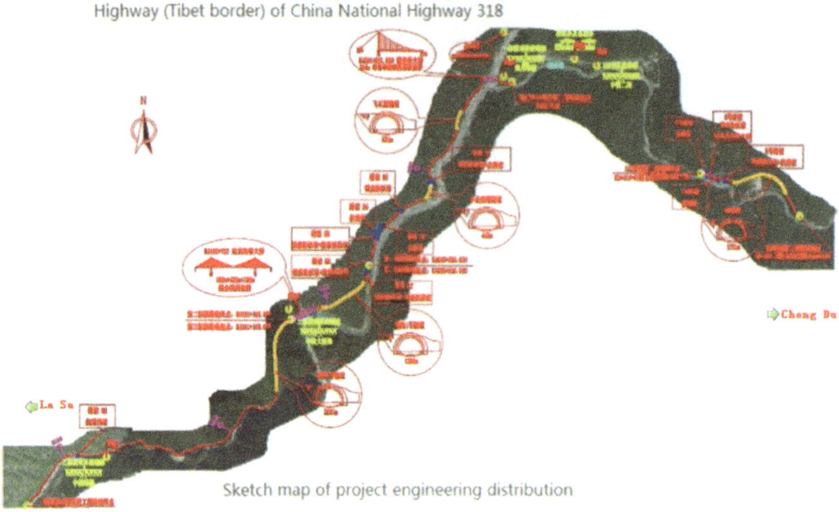


Sketch map of project engineering distribution

**Fig. 3.1** (continued)

**Fig. 3.1** (continued)

(c)

Fig. 3.1 (continued)

(d)

**Fig. 3.1** (continued)

driving. Figure 3.1c shows that the long-span bridge built in the Pailong section of the Sichuan–Tibet Highway crosses the Parlung Zangbo River, avoiding the phenomenon in which the former small-span and low-grade bridges could easily be destroyed. Figure 3.1d shows that before the construction of the tunnel or the shed tunnel in the Pailong section of the Sichuan–Tibet Highway, the highway spreads along the riverside of the mountain, which is narrow and dangerous to drive; after the construction, the road passes through the tunnel or shed tunnel,

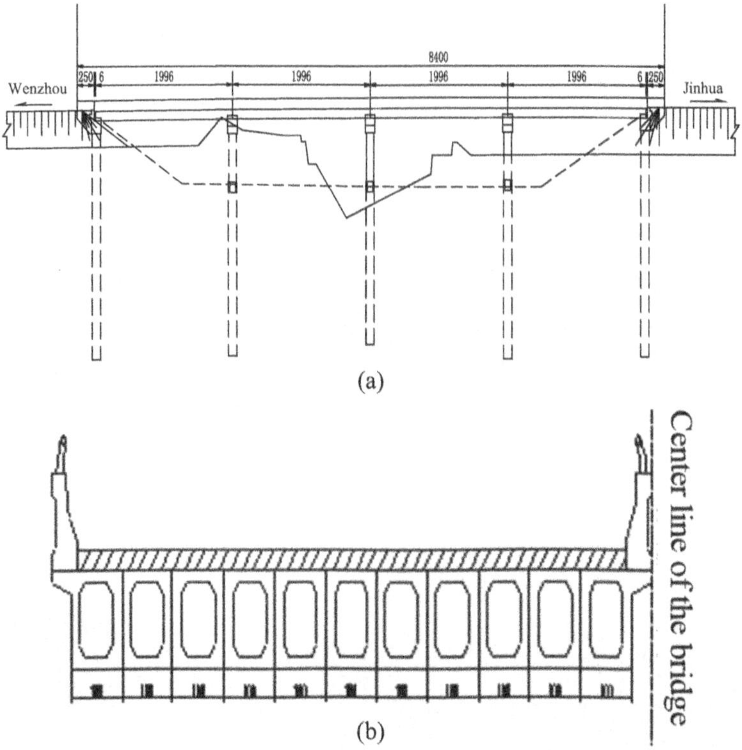


**Fig. 3.2** **a** Layout of the type of the bridge; **b** standard cross-section of the bridge

avoiding the narrow and dangerous riverside mountain line, so as to ensure the smooth flow of the traffic on the road.

(2) Analysis method (tactical problem): composite structure bearing force > external force, achieving a stable equilibrium

③ Calculate the equilibrium (stable equilibrium and control of the deformation coordination of engineering structures under stress and deformation are the key; before the 18th century, engineering structures had no mechanical calculations, so it is very important to learn from the classic engineering structures; mechanical equilibrium in engineering structural design and construction is only an auxiliary calculation and check.).

For example, the superstructure of a bridge uses the 4 × 20 m prestressed concrete simply-supported hollow slab, the bridge deck is continuous, and the substructure uses a cylindrical abutment and pile foundation. The layout of the type of the bridge is shown in Fig. 3.2a, and the standard cross-section of the bridge is shown in Fig. 3.2b. Large transportation vehicles have 30 axles, each of which is 20t, with a wheelbase of 1.5 m and a total weight of 600 t.

The specific reinforcement measures are as follows:

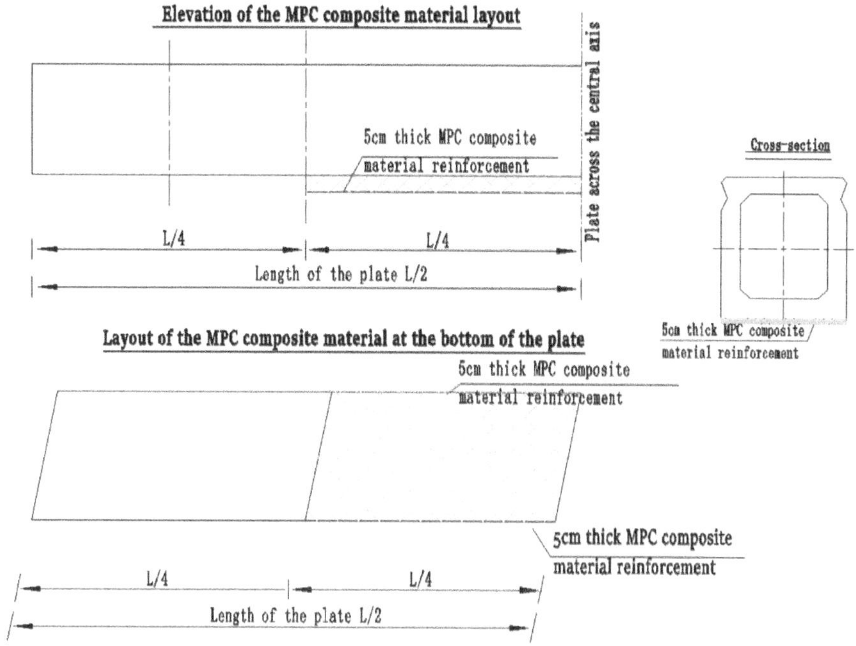


**Fig. 3.3** Structural diagram of a bridge reinforcement and reconstruction

(1) For the beams and slabs that do not meet the stress requirements in the transportation process, anti-breakaway anchor bolts should be installed at the bottom of the beam, and then 5 cm thick MPC composite material should be poured for reinforcement (Fig. 3.3);
(2) The bridge deck pavement should be removed and rebuilt. After the hinge joint of the beam and slab has been chiseled, the high-strength grouting material should be used for pouring again, and the planting reinforcement should be used for strengthening.

As shown in Fig. 3.4, the stress curve of the center line of the plate at the bottom of the beam along the direction of the bridge before and after the reinforcement is compared.

Under the action of a special load, the tensile stress at the bottom of the middle span of the beam slab structure is out of the limit before the reinforcement (as shown in the horizontal limit value in Fig. 3.4); for the beam slab of the middle span after the reinforcement, the principal tensile stress at the bottom of the beam under the action of a special load is smaller than that without reinforcement, and the tensile stress is less than the specified value.

In a word, the vertical stiffness of the 20 m simply-supported hollow slab girder structure is relatively small, which is only suitable for the vehicle action within the standard load of the volume of general traffic of an ordinary highway; it is not suitable for the overload action of a large volume of traffic or even part of the overload

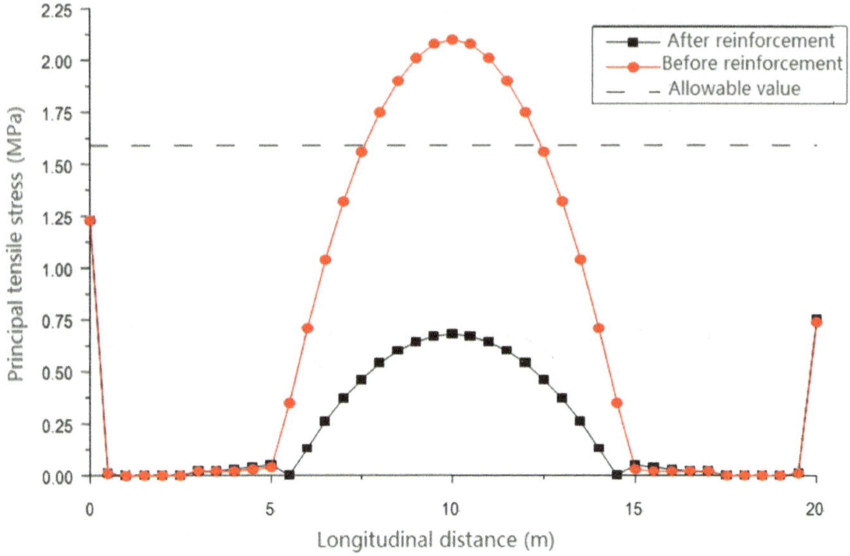


**Fig. 3.4** Comparison of principal tensile stress before and after the reinforcement

action of main highway bridges. This is consistent with the case in which the simply-supported hollow slab bridge beam of 20 m on the main road often bears heavy traffic or even partial overload, which can easily be damaged or broken or can even affect the operational safety. This is a case for adopting the analysis method to solve the specific problems of highway engineering structures.

# Chapter 4
# Simplification of Complex Problems

Modern engineering problems are becoming more and more complex. It is necessary to start from the engineering concept, grasp the main contradictions of the problem or the main aspects of the contradiction, establish a model reflecting the overall state of the stress, and then use modern mathematical and mechanical tools to analyze the engineering mechanical behavior under specific working conditions in detail, simplify the complex problems, and make sure that the results of the analysis cannot only meet the engineering accuracy requirements, but also reflect the mechanical law of the engineering system.

I. Simplification of complex problems is an important method of the mechanical analysis of engineering problems. In practice, we should distinguish between primary and secondary contradictions, grasp the main contradictions of the problem, and avoid the one-sided model of thinking. A classic example is the beam theory in mechanics of materials. Material mechanics is the theory of studying the state of the stress of material components such as bars, beams and columns. It is developed on the basis of solid mechanics by simplifying the problems through reasonable approximate assumptions. It has a good degree of accuracy for most engineering problems. For example, the Bernoulli beam theory greatly simplifies the mechanical calculation of the beam by introducing the assumption of the plane section, which not only avoids the complex mathematical formula, but also obtains the approximate solution meeting the engineering requirements, and realizes the purpose of simplifying the complex problem.
II. To simplify complex problems, we should unify the overall control with the details. From the perspective of dialectics, although the main contradiction dominates, it plays a leading and decisive role in solving problems, and the secondary contradiction is subordinate and plays a secondary role; there are differences and connections between them, which are interdependent and interacting. The main aspect of the contradiction determines the solution of the minor contradiction, which will in turn affect the solution of the major contradiction



H. Zhu and L. Shi, *Methodology of Highway Engineering Structural Design and Construction*, Advanced Topics in Science and Technology in China 59,
https://doi.org/10.1007/978-981-15-6544-1_4

if it is not handled well. For example, in order to build a tunnel in the diluvial body, the overall stability of the slope must be guaranteed first, which is the main contradiction of the problem, otherwise "A thing cannot exist without its foundation"; the stability of the surrounding rock of the tunnel is a secondary contradiction, but the control of the construction in the excavation process in turn affects the stability of the slope.

In fact, traditional mechanical knowledge, such as theoretical mechanics, material mechanics and structural mechanics, can be used to analyze the stress (internal force) and the state of the deformation of engineering structures, so as to grasp the main contradiction of engineering problems; while modern mechanics, such as continuum mechanics, is to solve the state of the stress (internal force) deformation of the structure by continuous integration from the stress-strain state of the particle, implicitly assuming that the medium particle is full of space, and the integration from the stress-strain state of the particle or the differential section to the state of the stress (internal force) deformation of the whole system is feasible, the result of the calculation is in equilibrium. If we consider the assumptions of "control the deformation coordination" and the "reasonable energy conversion" of the reasonable structure at the same time, and adopt modern mechanics, such as continuum mechanics, to solve the state of the deformation of the structural stress (internal force), the structure can always be made in a stable state of equilibrium. This is also the purpose of the "control of deformation coordination" and "reasonable energy conversion" of the system in the process of simplifying the complex problem of a reasonable structure and using the combination of traditional mechanics and modern mechanics to study engineering problems.

III. The simplification of complex problems is conditional, and the key is to keep the system stable and in equilibrium. A large number of practices have proved that the "control of deformation coordination" and "reasonable energy conversion" are the basic requirements for building a reasonable structural system and an analysis of the engineering structure. For example, most ancient buildings were built of wood or masonry materials. Although the strength of such materials is low and there are many joints between components, ancient craftsmen could choose the structural form with good geological conditions and reasonable stress according to local conditions, creating a batch of immortal excellent buildings. The Zhaozhou Bridge, which is more than 1,400 years old, is such a classic. These ancient buildings, under the guidance of simple mechanical thoughts, naturally meet the conditions of the control of "deformation coordination" and "reasonable energy conversion", which are the important reasons for their existence. In contrast, most modern buildings are built of steel and reinforced concrete materials, which very strong and many kinds of connective structures. But up to now, all kinds of engineering accidents have still occurred frequently. The important reason is that the stable equilibrium of the system is ignored and the basic requirements of the control of "deformation coordination" are not met. The modern theory of design mechanics, represented by static or dynamic continuum mechanics, studies the state of the stress (internal force) deformation of the structure. Only when the structure meets the conditions of control of "deformation coordination" and "reasonable energy conversion", can it meet the requirements of the initial design stress (internal force) deformation of the structure in the process of building construction or use, that is to say, to meet the reasonable transmission or transfer path of force. No matter whether it is surface engineering or the underground engineering, as long as the interaction with the rock and soil mass reaches a "stable equilibrium and control of the deformation coordination", the basic conditions for studying the "better solution" of the engineering structure can be expanded to "giving full play to the self-supporting capacity of the rock and soil mass" and "basically maintaining the original state of the rock and soil mass". Just like holding infants, when the holding method is different, the infants' stress and state of deformation is different. Adults in Yunnan or Guizhou use baby carriers to hold infants' buttocks, waist, neck and other pressure or bending parts, so the infants' stress and state of deformation is generally in a simple state similar to that of adults' stress and deformation, but it has nothing to do with the state of the movement of adults or the whole displacement of adults and infants. It can be seen that it is necessary to study not only the natural complex state of infants' stress and deformation, but also the simple and reasonable method of holding infants, so that the natural complex state of infants' stress and deformation can be transformed into a simple state similar to that of adults' stress and deformation. With the help of auxiliary carriers and other tools, professionals and general staff can hold infants, and keep infants' stress and deformation always in a normal state, which is a simple and reasonable approach.

The reconstruction and treatment of the wooden scaffold of the Sichuan–Tibet Highway landslide can provide an intuitive example of the stable equilibrium of the engineering. In the 1950s, the wooden scaffold was used to reinforce the landslide in the Tibet section of the Sichuan–Tibet Highway. After 50 years, the wooden scaffold had partially rotted, and a certain degree of compression deformation had occurred (Fig. 4.1). The landslide was in a critical state of equilibrium, which was an unstable equilibrium. There are two construction sequences for the reconstruction and treatment of these landslide sections: "first remove the wooden scaffold and then drive the anti-slide pile" and "first drive the anti-slide pile and then remove the wooden scaffold". There are different opinions on which scheme to adopt. Applying the idea of a stable equilibrium, the removal of the wooden scaffold will release the supporting force on the slope, which is a kind of disturbance to the slope. When the slope is in an unstable state of equilibrium, this kind of disturbance will lead to slope instability and a landslide accident. Therefore, although it is convenient for mechanical construction to remove the wooden scaffold first, the scheme of "first removing the wooden scaffold and then driving the anti-slide pile" should not be adopted. In contrast, using the scheme of "first driving the anti-slide pile and then removing the wooden scaffold" can basically keep the landslide body in the original critical state of equilibrium, with a large grasp and a small need for supporting force, and only the mechanical construction is not convenient.

**Fig. 4.1** Original state of the treatment of a landslide wooden scaffold in the Tibet section of the Sichuan–Tibet Highway

**Fig. 4.2** Landslide treatment of the slope supported by the former wooden scaffold in the Tibet section of the Sichuan–Tibet Highway

Figure 4.2 is the sequence of construction of "first driving the anti-slide pile and then removing the wooden scaffold", which has achieved a good effect of the engineering treatment. However, if the scheme of "first removing the wooden scaffold and then driving the anti-slide pile" is adopted, deformation and damage to the upper slope will inevitably be caused, which will lead to the deterioration of the mechanical properties of the rock and soil mass. It will not only threaten the safety construction of the anti-slide pile at the foot of the slope, but it will also require more anti-slide piles in order to resist the risk of the reduction of the anti-slide force caused by the deterioration of the rock and soil properties of the slope. This example embodies the guiding ideology of "basic maintenance of the original state of rock and soil mass", "coordinated control of deformation", "strong pre-supporting" measures and

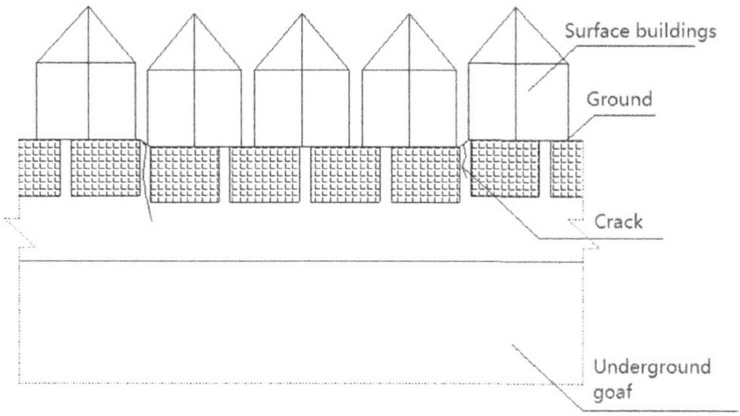


**Fig. 4.3** Diagram of how to protect surface buildings when conducting underground excavations

"reasonable energy conversion" proposed by the theory of equilibrium and stability. Only through the adjustment of the sequence of construction, can the effect of "skillfully deflecting the question" be produced.

IV. Analysis of typical cases. The application of the concept of simplified analysis of complex problems is illustrated with several engineering examples for readers' reference.

(1) How to protect surface buildings when conducting underground excavations

In the traditional calculation of the deformation of soil mechanics, soil is generally regarded as an elastic body, while in the stability calculation, soil is regarded as a rigid plastic body. In fact, the rock and soil mass is not an ideal elastic-plastic body, and its mechanical effect is not static, but related to time and process. Therefore, in a geotechnical engineering analysis, the simplified mechanical model is used to describe the complex engineering problems, and the engineering problems are transformed into mechanical and mathematical problems. The key is to make sure that the abstraction of the model is correct in engineering concepts and mechanical principles. A typical example, as shown in Fig. 4.3, is how to protect surface buildings (groups) when conducting underground excavations. Due to the special engineering size of long buildings (groups), it is difficult to ensure the coordinated deformation of buildings (groups) in the direction of the excavation. In excavation engineering, because of the uneven deformation of the ground, it is easy to cause damage to or even the collapse of the building, and because the scale of the underground rock and soil layer of the building is large in the direction of the excavation, using the traditional elastic-plastic model analysis will produce great error, that is, using the steady-state mechanical model to simulate the unsteady rock and soil mass will produce an error.

In order to simplify the problem and reduce the error caused by the simplification of the mechanical model, the error of analysis of each calculation unit is reduced by the method of "breaking up the whole into parts", and then the overall error is reduced, so as to achieve the purpose of controlling the coordinated deformation of the surface buildings. The specific method is as follows (Fig. 4.3):

① Add deformation joints to divide the long building into sections, and the length of each section should be controlled within 20 m, or the complex building should be divided into several parts, and the length of each part should also be controlled within 20 m.

② Additional measures such as a steel anchor tie rod, a steel light tie rod, a steel light ring beam or anchor plate should be presented to increase the rigidity of each section so as to resist the binding force.

Through the above measures, the use of deformation joints to control the coordinated deformation of surface buildings can effectively release the secondary stress caused by underground excavations and increase the system's ability to adjust for post-construction settlement; the scheme of sectional reinforcement can increase the stiffness of each section and resist the binding force of the building; however, the traditional mechanical method can still be used to analyze each section in order to realize the purpose of simplifying complex problems.

(2) Measures for a "lower partition and an upper sealing" of soft soil subgrade

Another typical case is the "lower partition and upper sealing" measures of soft soil subgrade. As for the specific design concept of a "lower partition and an upper sealing" of subgrade, no detailed introduction is made. Here, the measures are mainly introduced from the perspective of thoughts regarding the simplified analysis of complex problems. In the study of soft soil, the existing methods of consolidation and secondary consolidation are not perfect. Soft soil is a kind of three-phase system medium, with soil particles as the basic skeleton and water and gas filled in. Due to the interaction among three-phase media, the stress-strain relation is very complex. The deformation depends on the state of the stress when loading and is related to the loading history.

In terms of deformation, soft soil has consolidation and rheological properties under load. Consolidation is the process of excess pore water pressure dissipation, and rheology is the process of soil particle skeleton deformation. The two are paired up to produce the settlement of soft foundation. In the early stage of settlement, the deformation of soft soil is mainly caused by the joint action of consolidation and secondary consolidation rheology, mainly consolidation, supplemented by rheology. In the later stage of settlement, the soft foundation will further settle under the consolidation and secondary consolidation rheological actions, but the main one is rheology, and the secondary one is consolidation. Therefore, it is unreasonable to multiply the main consolidation settlement by the coefficient to calculate the total settlement. If neglecting consolidation and rheology, any property and its coupling effect will result in inaccurate results of the calculation.

From the analysis of stress, the soft soil subgrade is affected by a static load such as self weight and subgrade gravity and the dynamic load of the vehicle. Self weight and subgrade gravity belong to the semi-infinite linear load, that is, the uniform load with a stress boundary in the bridgehead direction and an infinite extension in the driving direction; vehicle load belongs to the point dynamic load, load amplitude changes according to the driving weight, and the load point position changes according to the driving speed. The proportion of the dynamic load in the total load changes according to the change in vehicle weight and subgrade height, so the whole stress situation is complex.

Considering the medium of transmission, the subgrade belongs to the soil-rock mixture structure. In the process of stress, the structure will change, and the transmission path will change. Therefore, the stress of soft soil is not the ideal continuous uniform stress.

It is very difficult to comprehensively consider the above-mentioned complex factors of deformation, stress and force transmission, and completely solve the settlement problem of subgrade from a mechanical point of view. In addition, even if the consolidation and secondary consolidation rheological properties of soft soil are considered, the existing methods of mechanical analysis often carry out an equivalent simulation by proposing the viscoelastic plastic model, which is generally created by the combination of spring, clay pot and slide block. In this process, it has been implicitly assumed that the soft soil foundation is in a state of equilibrium and stability, which is not fully consistent with the actual situation. At the same time, for the condition in which the simple soft soil medium belongs to the continuous medium, the "control of deformation coordination" is naturally satisfied, so the error generated by using the simple combination model of spring, clay pot and slide block is within the allowable range of engineering. This kind of model has been widely used in airport foundation treatment and has achieved a great deal of success. However, for the soft soil foundation at the bridgehead, the whole soft soil foundation is not a complete continuum due to the addition of long and short pile foundations or grouting measures to increase the strength. At this time, simply using the combined viscoelastic plastic model of spring and clay pot to carry out a mechanical analysis without considering the "coordinated control of deformation" will often produce a lot of errors.

In order to solve the above problems systematically by applying the principle of simplification of complex problems, the viscoelastic plastic model can be used to analyze the consolidation and rheological properties of soft soil foundation; then, on this basis, additional supports (such as long and short piles, buoyancy pipes, grouting, etc.) are added by following the principles of "control of deformation coordination" and "reasonable energy conversion", and the overall shape of the soft soil foundation is changed by a state of force. In this way, the main contradiction of solving the problem is transformed into how to use the model to consider the consolidation and rheological properties of soft soil. The two main aspects of the contradiction include:

① Dang-slag subgrade belongs to the mixture, which leads to the complexity of the stress mode of soft soil.

② The mechanical behavior of soft soil is complex, with properties of rheology and consolidation, and they are coupled with each other; the soft soil belongs to the unstable medium.

In view of the above two main contradictions: First of all, we can design through the measures of "lower partition and upper sealing". The upper sealing measures mainly control the erosion and corrosion of the subgrade caused by rainfall. The lower partition measures can not only prevent the infiltration of capillary water, but also mainly control the overall deformation of the subgrade, prevent local stress from being too great, and improve the overall ability to control the coordination of the deformation of the mixture subgrade, which can effectively reduce the influence of the rheological action of the soft soil. Second, we should reinforce the soft soil properly, improve the rheological threshold of the soft soil, and control the stress within the rheological threshold.

For the soft soil foundation with a large thickness and a poor quality of the soil, in order to effectively promote the "lower partition and upper sealing" measures for soft soil subgrade and other similar engineering measures as well as the development of new measures and new technologies, the key is the mechanical requirements of the overall stress and common deformation of the structure. It is impossible to accurately predict and control the deformation of the dense dispersion on the soft soil foundation. Although the subgrade is stable as a whole, it can be deformed locally, so the conventional mechanical calculation of the subgrade design will be deviated, which should be paid attention to.

It is advised to carry out the following three tasks in the design, construction and management of soft soil subgrade:

① According to the different load levels and local conditions, the effective measures of the subgrade lower partition meeting the mechanical requirements should be developed to ensure that the stress and state of the deformation of the subgrade are not affected by the rheological property of the soft soil foundation, which is conducive to the uniform and stable settlement of the subgrade and the reasonable stress of the pavement;

② A good upper sealing layer and drainage structure of the subgrade will make the subgrade dense and adhesive and not affected by moisture;

③ The consolidation and rheological properties of a soft soil foundation can be improved by appropriate treatment.

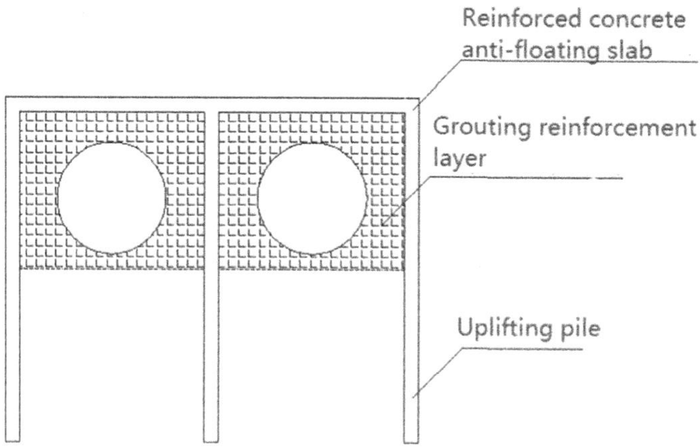


**Fig. 4.4** Diagram of the anti-floating design of pipeline under the river

The core of the above two cases is to adopt corresponding engineering measures and the concept of simplification of complex problems to transform the state of deformation of the complex stress (internal force) of engineering structures into the state of deformation of the simple stress (internal force) of engineering structures, so that the construction or use of complex engineering structures conforms to the initial state of deformation of the design stress (internal force) of the structure, and can ensure a stable equilibrium of the system.

(3) Anti-floating design of tunnel under the river
Tunnel work often needs to pass through muddy river beds. When the buried depth is insufficient and the thickness of the overlying soil is not large, it is easy for the floating phenomenon of the tunnel to occur due to the flow characteristics of the silt. If considering from a single tunnel, the systematic solution of such problems involves the rheological constitutive equation of silt, the interaction between the tunnel and silt, and the influence of tunnel spacing, etc., for which the analysis and calculation are very complex.
With the idea of simplification of complex problems and the concept of "coordinated deformation" and "reasonable energy conversion" of the system, engineering measures to prevent the tunnel from floating can be taken as shown in Fig. 4.4. By means of grouting reinforcement and an anti-floating reinforced concrete slab, the local deformation of a single tunnel is controlled to ensure the overall coordinated deformation of the tunnel system. At the same time, the uplifting pile is used to resist the buoyancy of the tunnel in the silt medium.

(4) Foundation treatment of a high-rise building above an underground cavern
Another case of simplification of complex problems is the foundation treatment of high-rise buildings above underground caverns. Under the core barrel of the building, at about 43 m away from the surface, and about

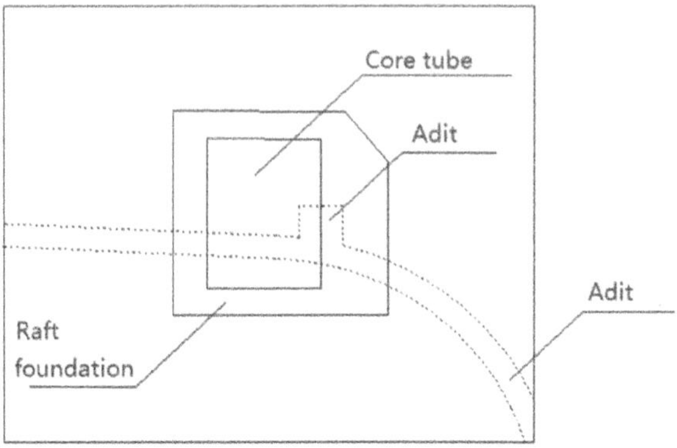


**Fig. 4.5** Diagram of the foundation treatment of a high-rise building above an underground cavern

25 m away from the floor of the 5th floor underground, there are air defense tunnels crossing the building. Because of the interaction between the foundation of the high-rise building and the lower cavern, it is easy to cause the instability of the upper building or damage to the lower cavern can easily occur. In order to effectively solve the engineering problems, the measures shown in Fig. 4.5 can be taken to deal with the foundation of high-rise buildings.

The raft foundation can be built around the core barrel, the upper structure thus adopts a sparse frame structure, the lower part of the core barrel adopts a short pile and thick raft foundation to support the load of the high-rise buildings, and the foundation outside the core barrel adopts artificial pile foundation to cross the rock stratum on the top of the air defense tunnel. In this way, a uniform stress of the foundation is ensured, the deformation of the whole system is coordinated and the energy conversion principle is met. Complex problems have been simplified.

(5) Bridge conceptual design

In the stage of bridge conceptual designing, the purpose of structural analysis is to ensure the establishment of the scheme. Therefore, the simplified analytical method (estimation) or finite element method of the simple model can be used. It mainly estimates the dead load (self weight) and live load (automobile, crowd) effects, and the live load can be simplified as the uniform load; during the estimation, it is necessary to simplify the calculation model of the structure as much as possible, plane the spatial problems, simplify the complex structure, and consider the influence of the construction process on the structural stress.

After the establishment of the structural scheme, in the mid-term stage of designing, according to the distribution of the actual load and the structural state, the finite element method should be used for an accurate calculation and analysis.

The core of the above three cases is to introduce the concepts of the "control of the deformation coordination" and the "reasonable energy conversion" under the guidance of the concept of the simplified analysis of complex problems, so as to facilitate the application of the combination of traditional mechanics and modern mechanics, and effectively solve the mechanical problems in the design and construction of complex engineering structures.

V. General idea of engineering design

Most modern buildings are made of steel and reinforced concrete, which are extremely strong and have many connecting structures. At present, the modern theory of design mechanics represented by static or dynamic continuum mechanics is mainly used for analysis and calculation. However, engineering safety accidents often occur during the construction or use of modern buildings. Analysis of this kind of "projects with problems" is the deep-seated cause of safety accidents, which is the state of deformation of the stress (internal force) in the process of construction or use that does not conform to the initial design of the structure. Therefore, the author thinks that it is an important basis for the rational structural design to ensure the "control of deformation coordination" and "reasonable energy conversion" of the building structure. The measures such as pre-support, reliable connection structure, test and inspection under complex geological conditions can correct the unreasonable design, guarantee the "coordinated control of deformation" and the "reasonable transformation of energy" of the building structure, and make sure that the structure conforms to the initial design stress (internal force) and state of deformation in the process of construction or use.

In the preliminary stage of the designing of modern buildings, traditional mechanical principles such as theoretical mechanics, material mechanics and structural mechanics can be used for design conception and estimation, so as to grasp the main aspects of the contradiction of a reasonable structure (Fig. 4.6). In the middle stage of design, advanced mechanical theories such as static or dynamic continuum mechanics can be applied to refine the calculation accurately, improve the auxiliary means and structural design, and realize the systematic analysis of the process of stress.

In this way, the reasonable structure can carry out the dynamic prediction and analysis of the stress (internal force) and deformation. Reasonable structural internal force, deformation and energy are three in one. Although only the deformation value of the structure can be obtained by detection, only under the condition that the internal force transfer or delivery and energy conversion of the structure are basically reasonable can the structural deformation have a rule to follow, and the structural deformation can be used to predict the state of the building's health. Therefore, only on the basis of the basic stable equilibrium of the engineering structure can we effectively predict the law of the engineering structure by using the mathematical model. To ensure the rationality of the engineering structure is the key, otherwise there will be meaningless results of the prediction.

In short, modern and future engineering research and practice cannot be separated from the experience and lessons of past engineering practice. On the

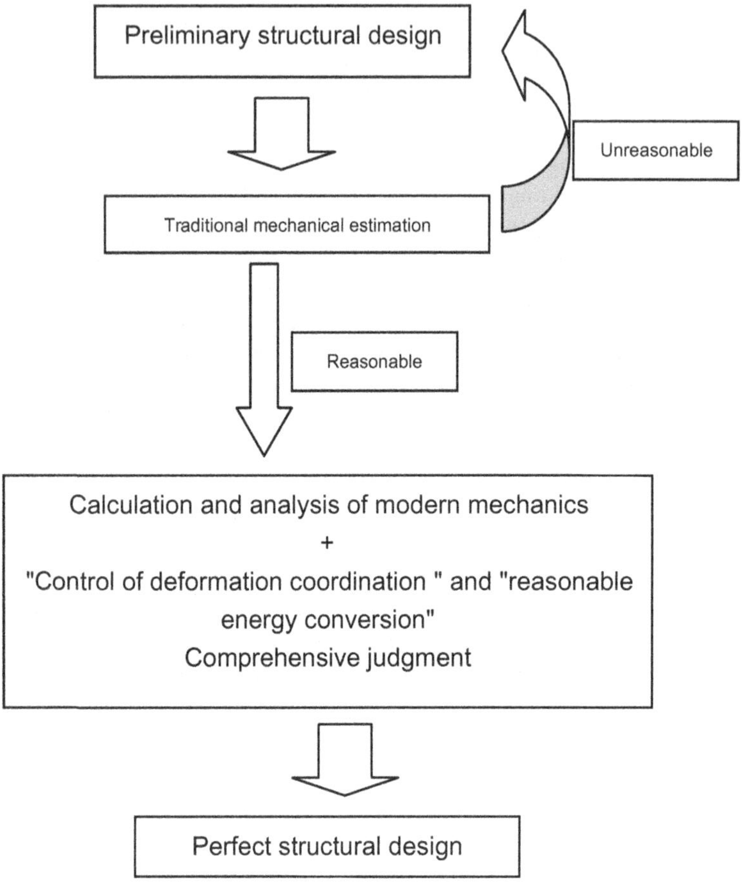


**Fig. 4.6** The process of the analysis of engineering problems

basis of previous working practice, the author summarizes the existing experience and lessons and traditional mechanical design methods, introduces the concept of the "control of deformation coordination" and "reasonable energy conversion" of engineering structures, and puts forward the theory of the equilibrium and stability of complex structures. In the process of preliminary calculation, it is necessary to take measures according to local conditions, grasp the main aspects of contradictions, analyze specific problems and simplify complex problems; then, on the basis of analysis and calculation, combined with engineering experience, introduce the basic requirements of the "control of deformation coordination" and "reasonable energy conversion" of the system for comprehensive judgment and preliminary engineering design; in the middle stage of design, advanced mechanical principles such as static or dynamic continuum mechanics are applied for detailed calculation and analysis, and auxiliary means and structural design are improved. In this way, the actual design and construction of the engineering structure should be based on the different stability of the state of equilibrium of the engineering structural system, and equilibrium, stable equilibrium, and stable equilibrium and control of deformation coordination should be used separately for analysis. The general situation (corresponding to the relatively simple mature engineering problems such as "apple landing site prediction") needs to consider the theoretical calculation and analysis; in special cases (corresponding to relatively new complex engineering problems such as the "prediction of leaf-falling site"), not only theoretical calculation and analysis, but also comprehensive methods such as a reasonable structural system, a reasonable construction method, an appropriate support, process control, etc. need to be considered, so as to truly achieve a "stable equilibrium and coordinated control of deformation".

# Chapter 5
# Application of Typical Engineering

## 5.1 Coordinated Control Method of Structural Deformation

To study $H_2O$ is to study the property or micro structure of water, but for shipbuilding we must study the potential of water, that is, the overall role of water; although both are to study water, the property and the potential of water are not the same; only studying $H_2O$ cannot complete the shipbuilding business, and the potential of water is not completely contained in $H_2O$; and it is the same for calculus in engineering mathematics or mechanics. The implicit assumption of deformation coordination in $\Sigma X_i$ is that although the integral is a differential sum, it is also conditional, that is, the problem of the control of deformation coordination.

In the calculation of the stress and deformation of engineering structures, there is an implicit assumption of deformation coordination, while some structures are in the same state of uncertainty or instability as infants or young children, so it is difficult to set specific measures in the methods and software of their calculation. At present, when the results of the calculation of engineering structures are inconsistent with the actual results, in some cases, the parameters of the calculation are often adjusted or the model of calculation is modified so that the calculation results are consistent with the actual results, while the coordinated control of structural deformation is ignored, which may lead to problems of the quality and even of the safety of engineering structures. Therefore, the measures for the control of deformation coordination must be adopted in the design and construction of engineering structures. Only by improving the structural design and construction scheme can the deformation coordination conditions be met, and the results of the calculation of engineering structures can be close to the actual results, so as to ensure the quality and safety of the engineering structures. For example, it is generally accepted in the world that the calculation of soft soil foundation force is more accurate and the calculation error of deformation is larger. The settlement calculation of soft soil foundation mainly reflects the trend. The settlement of soft soil subgrade has to rely on control measures.



H. Zhu and L. Shi, *Methodology of Highway Engineering Structural Design and Construction*, Advanced Topics in Science and Technology in China 59,
https://doi.org/10.1007/978-981-15-6544-1_5

The pile plate structure (patent of China Railway Siyuan) is used in high-speed railways to control the settlement of soft soil after construction, and lime soil subgrade (patent of Jiangsu Province) is used in high-grade highways to control the settlement of soft soil after construction. After the settlement of the soil subgrade after construction, lower partition + pile transition (patent of Zhejiang Province) is adopted for high-grade highways to control the post-construction settlement of the soft soil slag subgrade.

Traditional Chinese medicine diagnoses diseases by means of looking, listening, questioning and feeling the pulse, that is, the connection or main aspects of diseases; Western Medicine finds out diseases by means of inspection via equipment, that is, the symptoms or specific aspects of diseases; therefore, in order to truly grasp the causes of diseases and achieve the purpose of the cure, we should not only grasp the main contradictions but also analyze the specific problems. The same is true for the engineering structure. It is necessary to grasp the problem of the coordinated control of structural deformation as well as carry out an accurate analysis of the structure, so as to achieve the unity of an overall control of the structure and an accurate analysis of the structure, and ensure the quality and safety of the engineering structure.

I. Physical significance of the control of the deformation coordination of the engineering structure

The influence of the control of deformation coordination on the stability of the state of structural equilibrium is intuitively understood through the diagram of stable equilibrium and the control of the deformation coordination of heavy objects shown in Fig. 5.1. The weight W in Fig. 5.1 is suspended by n-strand rope, and the rope force $P_1$, $P_2$,…,$P_n$ and the self weight W are in equilibrium. The state of equilibrium is different because the relationship of the control of deformation coordination of the joint action of the rope force $P_i$ and the weight W is different, and the state of equilibrium stability is also different, which is shown as follows:

(1) When the weight W is under the static load, $P_1$, $P_2$,…,$P_N$ and W are in equilibrium. When $P_1$, $P_2$,…,$P_n$ are in stable equilibrium within the allowable range of their strength; when a $P_i$ exceeds the limit and breaks, the remaining n–1 cables will be redistributed. In the process of the redistribution of the internal force, there may be two situations: If the internal force can be transferred reasonably, the remaining n–1 rope is still in the range of strength, and the system will be in equilibrium again; if the structure of the system is designed improperly and the internal force of the system cannot be transferred reasonably, the internal force of the remaining n–1 rope will exceed the range of strength again and cause fracture. The repeated occurrence of this process will cause a chain reaction and make the whole system unstable. From the point of view of energy transfer, the above phenomenon can be explained as follows: due to the effect of weight W, the accumulated strain energy in each rope is $U_i$ ($i = 1,2,\ldots,n$), the structure is in stable equilibrium. If the stored energy $U_i$ in a rope reaches its energy absorption limit and breaks, $U_i$ will be

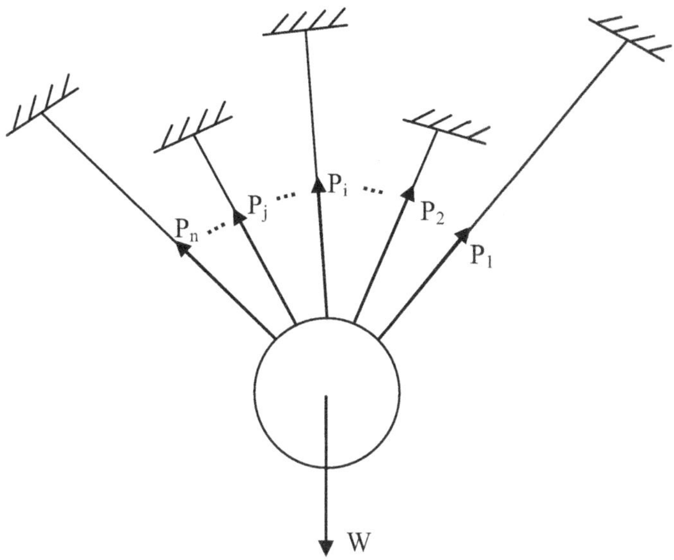


**Fig. 5.1** Diagram of the relationship between stable equilibrium and control of the deformation coordination of heavy objects under multi-strand rope suspension

released completely. As the total energy of the system remains unchanged, the energy of the structural deformation will be redistributed, and there will be two situations: if the energy can be reasonably transferred, the remaining n−1 rope can effectively absorb all the deformation energy, and the system will be in equilibrium again; if the structure is not designed properly, the work of the external force will break through the structural energy storage limit again, resulting in the fracture of the remaining n−1 rope, causing a chain reaction and the overall instability of the system.

(2) When the weight W is disturbed, the weight W will deviate from the original position, so $P_1$, $P_2$,…,$P_n$ will be redistributed only if there is $P_1$, $P_2$,…,$P_n$ deformation coordination (i.e. P1, P2,…,$P_n$) When the internal forces between PN can be reasonably transferred, and all of them are within the allowable range of strength, the system will be in a stable equilibrium and the control of deformation coordination, so that the system can recover its original position. Otherwise, when the structural design is unreasonable, the internal forces of the system cannot be transferred reasonably. The redistribution of P1, P2…,$P_n$ may cause a rope force pi to exceed the limit and even lead to a chain reaction and overall instability of the system. From the perspective of energy analysis, it is better to understand that the external disturbance to the weight W will input a certain amount of energy to the structure. Only when the structural deformation is coordinated (that is, the overall deformation energy can be reasonably transferred between each rope), and considering the energy consumption mechanism (such as resistance, friction,

**Table 5.1** Relationship between the control of deformation coordination and the state of structural equilibrium

| Contents / State | Force bearing analysis of the engineering structure | Applicable conditions |
|---|---|---|
| ① Stable equilibrium | The precise method of analysis is used to solve the problems of the engineering structure | Implicit or natural satisfying control of deformation coordination |
| ② Stable equilibrium and control of deformation coordination | First, the whole control and details are grasped, and then the precise method of analysis is used to solve the problems of the engineering structure | Build a reasonable structural system, reasonable construction method or process and effective process control measures to ensure the reasonable transmission or transfer path of the force |

etc.) of the air and the structure, can the system recover its original position. Otherwise, when the structural design is not reasonable, the internal energy of the system cannot be reasonably transferred. The redistribution of $U_1$, $U_2 \ldots, U_n$ may result in the failure of a rope to absorb the energy it should have and even a chain reaction leading to the overall instability of the system. In the analysis of general engineering structural mechanics, load and deformation adaptability are implicit. In fact, the stable equilibrium and the control of the deformation coordination of engineering structures are coupled. In practice, we often neglect it!

In order to better understand the internal relationship between the two methods of stable equilibrium and stable equilibrium and control of the deformation coordination of engineering structures, the problems of stress analysis and the conditions of the theoretical application of engineering structures corresponding to these two concepts are summarized as shown in Table 5.1.

Therefore, the stable equilibrium and the control of the deformation coordination of engineering structures are the unity of the mutual coupling of engineering structural problems, in which a stable equilibrium involves the load influence, and the control of the deformation coordination involves structural rationality. Although the two solutions are different, only the engineering structure which meets the dual mechanical and deformation control conditions of the structural stable equilibrium and the control deformation coordination is safe. Just as diseases are accompanied by each other, the problem of diseases can only be treated in the same way as solving diseases. However, if the problem of diseases is solved in the way that diseases are treated, it can never be eradicated, and vice versa.

II. The relationship between the metastable equilibrium of engineering structures and engineering accidents is revealed

In the past, based on step-by-step investigation and continuous improvement, it is easy to ignore the differences between "adjacent generations" of structures in a large number of engineering structural cases, and it is difficult to find the correlation between the metastable equilibrium of engineering structures and engineering accidents. From 2005 to 2015, the author counted about 40,000 successful and invalid engineering cases of the Zhejiang highway system, bridges, tunnels, soft soil subgrade, etc. at home and abroad. It is easy to find the differences between "alternate generations" of structures through the overall investigation and concentrated comparative study of these cases. It was found that all civil engineering structures satisfying the coordinated control of stable equilibrium and deformation are safe; on the contrary, there are some problems or even safety risks in varying degrees, revealing the correlation between the metastable equilibrium of highway engineering structures and engineering accidents. For example: People's back pain is not necessarily due to a problem of hunchback, and hunchback people do not necessarily have back pain; some engineering problems of "unclear, unclear road" have more geotechnical aspects and many structural aspects, which do not conform to the hypothesis or theory of deformation coordination. This kind of engineering cannot completely solve the problem only by traditional mechanics, but it can be solved by the method of overall control and a grasp of the details. The common problem of structural deformation coordination needs human control. When the engineering structure does not meet the control of deformation coordination, although the process of calculation is correct, the results of that calculation are different from the actual results, and there is no practical significance!

The existing theory of equilibrium and stability of engineering structures has had sufficient research on equilibrium, which mainly focuses on the theory of deformation coordination, the hypothesis of deformation coordination, etc., while there is less systematic research on the relationship between reasonable structures and mechanics for the control technology of deformation coordination, and it is difficult for some structures to guarantee the adaptability of structural load medium and avoid the problem of structural metastable equilibrium. The stable equilibrium and the control of deformation coordination of engineering structures are the unity of the coupling of the problems of engineering structures, in which the stable equilibrium involves the influence of the load while the control of deformation coordination is related to structural rationality. In the process of research, the author realized that engineering mechanics is based on the determination of material properties and micro structures, but some engineering structures will change in the process of material properties and micro structures, and the manner and law of that change are unknown. For example, in the process of the stress of soft soil materials of the soft soil subgrade, the results of SEM scanning of the micro structure show that the structure of soft soil particles changes with the stress The degree varies with the above requirements of engineering mechanics. The application of engineering mechanics to the analysis of the mechanical deformation of engineering structures sometimes results in deviation or even harmful results. Therefore, the design of engineering structures must meet the control conditions of deformation coordination, so that the engineering structure is always under normal stress and a state of continuous deformation. Otherwise, it will produce an unreasonable transfer of force and form a new or even harmful state of equilibrium, which meets the causal law of engineering mechanics with difficulty. How can we solve the problems that cannot be completely solved only by engineering mechanics? Although the traditional theory of the prediction of "stable structure" is the same as that of "metastable structure", but the actual methods of prediction are different; however, it is easiest to solve "metastable structure" by using the method of maintaining the stability of the movement of people (such as young people, adults, the elderly and so on). Based on the analysis of the stable equilibrium of the traditional engineering structure, the coordinated control technology of the deformation of the highway engineering structure is proposed for the first time at home and abroad. To solve the relatively complex engineering problems of the "metastable structure", the method of overall control and detailed control should be adopted comprehensively, that is, when the engineering structure is in the metastable state, the method of the control of the engineering structure should be adopted In order to control the stability of the engineering structure in the state of stress and deformation, a reasonable structure should be designed according to the standard of construction measures, analogy and test determination, and the "metastable structure" should be transformed into a stable structure first, then the precise analysis method should be used to solve the engineering problems. For example, the crack treatment method of highway soft soil subgrade is to improve the design method of soft soil subgrade on the basis of controlling the stress and deformation stability of soft soil subgrade by adding a lower partition or frame at the base of the improved soil or slag road: A. When the lower limit of soft soil rheology is higher than the stress value at the base of the soft soil road, the post-construction settlement of the soft soil subgrade can be controlled; B. When the lower limit of soft soil rheology is higher than the stress value at the base of the soft soil road, the post-construction settlement of the soft soil subgrade can be controlled. If the threshold value is less than the stress value at the base of the soft soil road, engineering measures should be taken to control the settlement of the soft soil subgrade within the allowable range after construction, so as to control the cracking of the road surface.

In the past, the methods of designing highway bridges, tunnels and soft soil subgrade all had some problems, for which it was difficult to guarantee the adaptability of structural load-carrying medium and avoid the metastable equilibrium of the structure. There are many successful examples and some unsuccessful examples in the cracking of highway bridges, tunnels and the soft soil subgrade pavement, for which it is difficult to solve by methods of engineering mechanics, either classified as construction problems, or there is no other way. On the basis of the existing design specifications, the methods of designing methods highway bridges, tunnels and soft soil subgrade are modified according to the control technology of the deformation coordination of traffic structures, which ensures the adaptability of the medium of structural transmission and avoids the problem of structural metastable equilibrium.

(1) Improve the method of designing highway bridges: In the past, under the repeated overweight load, the non elastic parts of small and medium span bridges and some special components of bridges would accelerate the accumulation of damage, especially in the 1980s, and most of the optimized bridges had problems. The previous method of designing had difficulty in avoiding such problems, which was also one of the reasons that overload easily led to the collapse of highway bridges. The core of the problem was that it was difficult to guarantee the adaptability of the structural transmission medium and avoid the metastable equilibrium of the structure. Bridge design should be improved by adopting the control technology of the deformation coordination of traffic structures. On the basis of the design specifications, the stiffness of small and medium-span highway bridges should be moderately improved, and the deformation coordination of some special components of the bridge should also be perfected. Under the repeated overweight load, the accumulated damage of the inelastic parts will be in a controllable range, avoiding the problem for which the appropriate overload will easily lead to the collapse of the highway bridge. It can ensure the adaptability of the structural transmission medium, avoid the problem of structural metastable equilibrium, and ensure the safety of the engineering structure. For example: a method for the reconstruction of the bridge deck of the floating tied arch bridge, aiming at the problem in which the bridge deck system of the bridge deck floating tied arch bridge has a large longitudinal and transversal deformation, and the suspender (especially the short suspender) steel wire can easily produce fatigue damage, a new technology of tied arch bridge reconstruction was proposed which can effectively control the structural deformation and improve the integrity; the key is to control the limiting structure of the bridge deck longitudinal and transversal drift;
(2) Improve the method of designing the highway tunnel: The New Austrian Tunneling Method (NATM) originates from hard rock, which has poor adaptability to soft rock and broken rock and soil mass, and is prone to mutation when encountering water. Therefore, the deformation control measures of rock and soil mass must be adopted. The theory of equilibrium and stability and key technology of underground engineering structures have been established. Based on the whole equilibrium and stability of the surrounding rock and supporting structure, the New Austrian Tunneling Method (NATM), the Norwegian Tunneling Method (NTM) and the New Italian Tunneling Method (NITM) have been classified and are compatible. According to different stages of the bearing capacity of the surrounding

rock, the existing achievements can be used to prevent and control the quality and safety of engineering; according to the control technology for the deformation coordination of traffic structures, the design of tunnel construction has been improved. On the basis of the design and construction specifications, according to the different stages of the bearing capacity of the surrounding rock, the existing achievements can be used to prevent and control the quality and safety of the project, to ensure that the force is transferred according to the design path, the unreasonable or even harmful transfer of the control force, and to avoid the sudden change or even collapse of the surrounding rock. For example, there is a kind of control technology for the deformation coordination regarding the construction of a soft flow plastic soil tunnel. In view of the fact that the existing new tunnel part needs to be buried at a shallow depth and concealed excavation should pass through the soft flow plastic soil stratum, the shotcrete anchor support technology cannot control the surrounding deformation and ground settlement of the tunnel at all during the construction. According to the control technology of the deformation coordination of traffic structures, the design and construction have been improved, and it is advantageous to pay attention to the reasonable excavation and support method of the tunnel. In order to control the plastic deformation or loose deformation of the surrounding rock, ensure that the force is transferred according to the design path, control the unreasonable or even harmful transfer of the force, and avoid the sudden change or even collapse of the surrounding rock. In other words, it is suggested adopting the surrounding support of steel gusset plate in advanced pipe shed limit, grouting on the face of the tunnel, improvement of the excavation support method, steel arch frame and shotcrete bearing all initial support loads, or horizontal rotary spray. The pile advance support, forming the supporting structure around the steel gusset plate in advance of the pipe shed limit, can better meet the stability of the surrounding tunnel and even the tunnel face during the construction process.

(3) Improved method of designing the highway soft soil subgrade: Long structures such as the highway soft soil subgrade are plane strain problems, which have a great influence on soft soil rheology. In the previous design, the influence of soft soil consolidation and secondary consolidation was taken into account, especially the lack of integrity of the subgrade such as soil-rock mixture, which makes it difficult to overcome the influence of soft soil rheology and is therefore prone to the post-construction settlement of soft soil subgrade, and prone to the problem of longitudinal and transversal cracking of the pavement. For the first time at home and abroad, the lower threshold of soft soil rheology was used to control the post-construction settlement of the soft soil subgrade. According to the control technology of the deformation coordination of traffic structures, the method of designing the soft soil subgrade was improved. On the basis of the design specification, the integrity of the stress and the state of deformation of the subgrade is first paid attention to, and then (A) when the lower threshold of soft soil

rheology is higher than the stress value of the soft soil subgrade base, the post-construction settlement of the soft soil subgrade can be controlled; (B) when the lower threshold of soft soil rheology is less than the stress value of the the soft soil subgrade, engineering measures need to be taken to control the settlement of the soft soil subgrade within the allowable range after construction, which is conducive to solving the problem of longitudinal and transversal cracks in the surface of the road. For example, a control method for the post-construction settlement of highway soft soil subgrade, aiming at the problem of the integrity of the stress and the state of the deformation of the existing newly-built slag subgrade filled with soil-rock mixture, but the integrity of the stress and the state of the deformation of the newly built slag subgrade filled with soil-rock mixture is controlled by the measures of a frame between the bottom layer of the subgrade and the top layer of the soft soil foundation, so as to avoid the problem of longitudinal and transversal cracking of the highway pavement. Since its application in 2000, highway bridges, tunnels, soft soil subgrade, etc. have been successful cases, solving the problems in the methods of designing highway bridges, tunnels and soft soil subgrade, as shown in Table 5.2.

**Problem 5.1.1**
The original Tongmai suspension bridge is a temporary project with guaranteed accessibility. As shown in Fig. 5.2a, the bridge is designed as a double tower and double span suspension bridge with a main span of 210 m and a load capacity of 20 tons. The main reason for the damage is that the main beam is composed of a Bailey steel frame beam connected at the bottom, and not only is the vertical rigidity not enough, but also the transversal rigidity and torsional rigidity are not enough. The original design probably focused more on the vertical load effect, less on the greater secondary stress of the main beam caused by insufficient rigidity, so the greater secondary stress of the connection between the suspender and the main beam caused the effect of cumulative damage to the main beam and the suspender and other components (such damage occurs many times in the tied arch bridge of the highway).

**Solution** For suspension bridges in mountainous areas with a great risk of lateral uneven deformation, a reinforcement method is proposed to effectively control structural deformation coordination and improve the overall stability of the structure. For medium and small span bridges ($L < 300$ m), as shown in Fig. 5.2b, c, the above problems can be solved by tensioning two arc-shaped transverse oblique limiting cables on both sides of the bridge and connecting the limiting cables with the main girder of the bridge by means of short cables. The two ends of the limiting cables are anchored on the bank or mountain on both sides of the bridge; in the actual project, four tunnels and two bridges were used to prevent collapse, water damage and other disasters, as shown in Fig. 5.2d.

**Table 5.2** Comparison of the main research results with similar research at home and abroad

| | Similar technologies at home and abroad | Technology of the project |
|---|---|---|
| The relationship between the metastable equilibrium of engineering structures and engineering accidents | In the past, it was easy to ignore the differences between "adjacent generations" of structures and find the relationship between the metastable equilibrium and uncoordinated deformation of engineering structures and engineering accidents | In the past 10 years, the author has counted about 40,000 cases of success and failure within the Zhejiang highway system, bridges, tunnels, soft soil subgrade and other projects at home and abroad, and made an overall investigation of these cases, a concentrated comparative study, and it was easy to find the differences between "alternate generations" of structures; it was found that all civil engineering structures that meet the condition of stable equilibrium and the control of deformation coordination are safe; on the contrary, there are some problems and even security risks to different degrees, and the correlation between them was found |
| Coordinated control technology of structural deformation in highway engineering | The existing theory of equilibrium and stability of engineering structures has sufficient research on equilibrium, which mainly focuses on the theory of deformation coordination, the hypothesis of deformation coordination, etc., while there is less control technology for deformation coordination for systematic research on the relationship between reasonable structures and mechanics, and it is difficult for some structures to guarantee the adaptability of structural load medium and avoid the problem of structural metastable equilibrium. The existing theory if equilibrium and stability of engineering structures has sufficient research on equilibrium, which mainly focuses on the theory of deformation coordination, the hypothesis of deformation coordination, etc., while there is less control technology for deformation coordination for systematic research on the relationship between reasonable structures and mechanics, and it is difficult for some structures to guarantee the adaptability of the structural load medium and avoid the problem of structural metastable equilibrium | Engineering mechanics is based on the determination of material properties and micro structures, but the material properties and micro structures of some engineering structures will change while under stress, and the manner and law of that change are unknown; the project first proposed the coordinated control technology of the deformation of highway engineering structures at home and abroad, on the basis of traditional analysis of the equilibrium and stability of engineering structures, engineering structural design and construction. The mechanical control condition of structural stable equilibrium was changed into the double control condition of mechanics and deformation of structural stable equilibrium and the control of structural deformation coordination, so as to avoid the inconsistency between the design state of stress and deformation of some engineering structures and the actual state of stress and deformation, to ensure the adaptability of the structural transmission medium, to avoid the problem of structural metastable equilibrium, and to ensure the safety of the engineering structures |

(continued)

**Table 5.2** (continued)

| | Similar technologies at home and abroad | Technology of the project |
|---|---|---|
| Improvement of the method of designing highway bridges, tunnels and soft soil subgrades | In the past, the methods of designing highway bridges, tunnels and soft soil subgrades all had some problems, for which it was difficult to guarantee the adaptability of structural load-carrying medium and avoid the metastable equilibrium of the structure. There are many successful examples and some unsuccessful examples in highway bridges, tunnels, soft soil subgrades and other problems, which are difficult to solve by using the existing methods of engineering mechanics, or they are classified as construction problems, or there is no way | On the basis of the existing design specifications, the design methods of highway bridges, tunnels and soft soil subgrades have been improved according to the control technology of the deformation coordination of traffic structures; the theory of equilibrium and stability and key technology of underground engineering structures have been established. On the basis of the overall equilibrium and stability of surrounding rock and supporting structure, the New Austrian Tunneling Method (NATM), the Norwegian Method of Tunneling (NTM) and the New Italian Tunneling Method (NITM) have been classified and are compatible, which can be based on the bearing capacity of the surrounding rock at different stages. It is the first time at home and abroad that the lower threshold of soft soil rheology has been used to control the post-construction settlement of soft soil subgrades. A series of patent technologies have been effectively developed to ensure the correctness of the improved method of designing. Since 2000, highway bridges, tunnels, soft soil subgrades and other successful cases have solved the design problems of highway bridges, tunnels and soft soil subgrades |

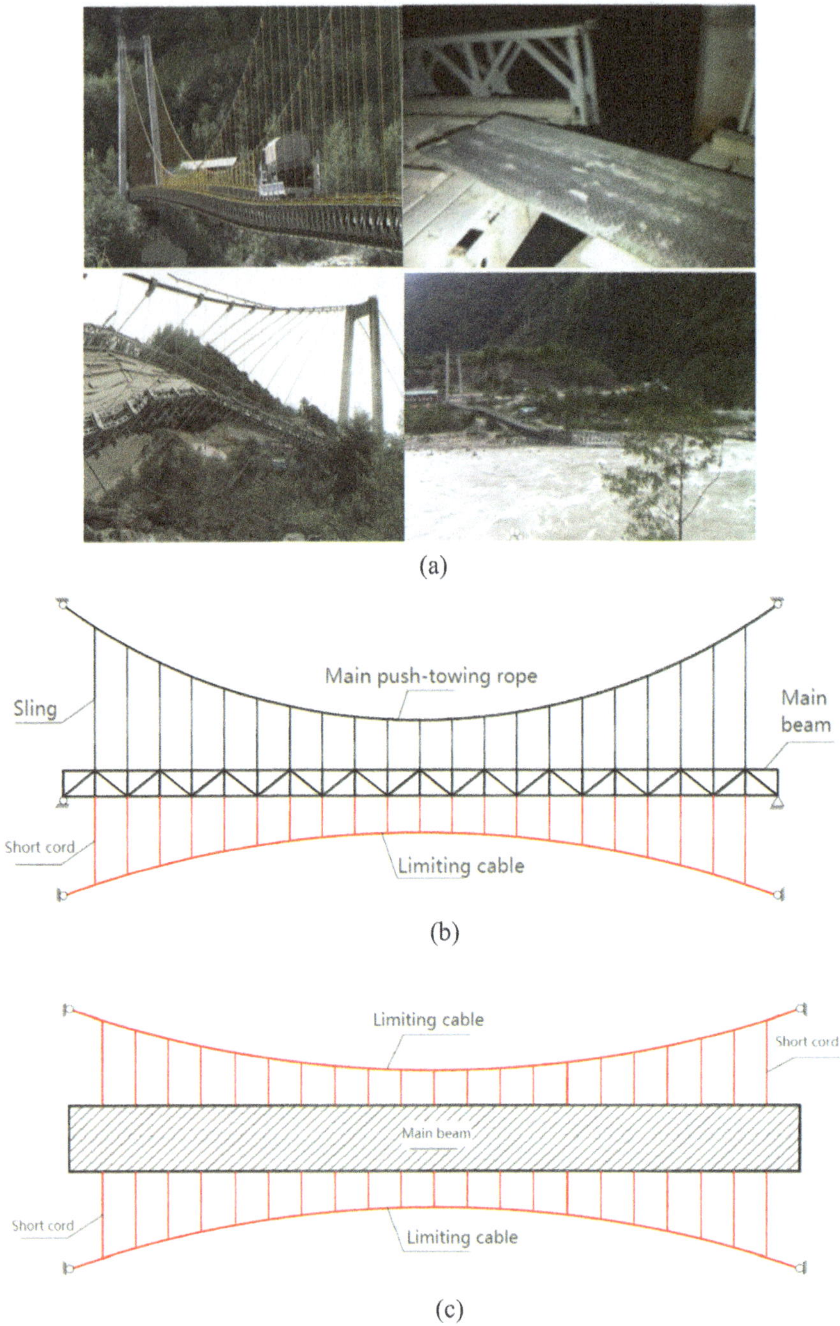


**Fig. 5.2** **a** Status of the original Tongmai suspension bridge before and after the damage; **b** Diagram of the vertical layout of the limiting cable of the medium and small span suspension bridge; **c** Diagram of the layout plan of the limiting cable of the medium and small span suspension bridge; **d** Comparison of three bridges before and after the Tongmai Bridge

(d)

**Fig. 5.2** (continued)

**Problem 5.1.2**

The middle through arch bridge has been widely used in recent decades due to its beautiful shape and large span capacity. However, the problems caused by the durability and reliability of suspenders have become increasingly prominent, and this has become a universal problem that cannot be ignored. For example, the collapse of the bridge deck of the short suspender fracture part of the Xiaonanmen Jinshajiang Bridge on November 7,2001, the collapse of the bridge deck of the short suspender fracture part of the Peacock River Bridge on national highway 314 on April 12, 2011, the collapse of the bridge deck of the short suspender fracture part of the Fujian Gongguan Bridge on July 15, 2011, etc., as shown in Fig. 5.3a; in these examples, the short suspender not only bears the tension (included in the calculation), but also bears the shear force of the bending (not included in the calculation), the accumulated damage to the structure is produced, and the comparison of the scanning test results of the short suspender steel wire shown in Fig. 5.3b is the proof. This type of bridge is a half-through or through tied arch bridge, and its deck system belongs to the floating system, as shown in Fig. 5.3c, that is, the system of the bridge deck structure with a suspender system crossbeam and a longitudinal bridge deck laid on the crossbeam is supported by a single suspender. In addition, the anchorage end of the suspender of the bridge deck system has severely rusted, both of which will affect the service life of the structure. Once a suspender fails, the bridge deck structure becomes a variable system, the bridge decks of two adjacent spans of the suspender will inevitably fall, resulting in the collapse of part of the bridge decks of some through tied arch bridges due to the sudden fracture of the suspender.

**Solution** Aiming at the problem in which the longitudinal and transversal deformation of the bridge deck system of the floating tied arch bridge is large, and the steel wire of the suspender (especially the short suspender) is prone to fatigue damage, a new technology for the reconstruction of the tied arch bridge could be proposed, which could effectively control the structural deformation and improve the integrity, as shown in Fig. 5.3d. The main measures include: (1) Pouring materials at the end of the bridge deck at the cross beam of the arch bridge by means of chiseling. Steel bars are planted on the top surface of the beam to hook the bridge deck, and composite materials of a high degree of toughness such as MPC fill in the gap between the beam ends to connect the bridge deck and the transverse beam as shown in Fig. 5.3e; (2) Longitudinal and oblique steel beams are added between the beam and the arch rib of the shortest suspender to limit the longitudinal and transversal displacement of the bridge deck, restore the characteristics of the design stress deformation of the short suspender and prevent the short suspender from breaking, as shown in Fig. 5.3f; (3) In order to meet the structural deformation requirements such as the effect of temperature change, expansion joints are set in the middle of the bridge deck system, as shown in Fig. 5.3e.

**Problem 5.1.3**

There are many layered phenomena in the soft soil layer of coastal cities, especially the problem of the shield construction for the subway crossing under the lower-soft and upper-hard soft soil layer being prone to floating up. When the layered combination of the soft soil layer is unfavorable, it is difficult to prevent the shield from floating up through the lead block weight pressing, which affects the segment installation and the longitudinal line position of the subway.

However, the existing methods of shield construction for the subway crossing under the lower-soft and upper-hard soft soil layer easily ignore the overall mechanical analysis of the interaction between the stratum and the shield, and there can be a lack of macro qualitative control methods. For example, in the process of interval propulsion of a medium depth EPB shield, the overall attitude of the shield machine floats up, and the track of its moving axis is shown in Fig. 5.4a. When the shield floats, the upper part of the shield driving cross-section is relatively hard sandy silt mixed with silt ③ 61 (standard penetration 14), sandy silt muddy silty clay ③ 7 (standard penetration 4), and the lower part is relatively soft sandy silt muddy silty clay ③ 7 (standard penetration 4), muddy silty clay ⑥ 1 (standard penetration 2). The shield machine is a Φ 6340 mm earth pressure equilibrium shield with mud, with a shield length of 8.0 m and a buried depth of 12 m at the top of the shield.

**Solution** In view of the existing method of the treatment of shield construction for the subway crossing under the lower-soft and upper-hard soft soil layer, it is easy to ignore the overall mechanical analysis of the interaction between the stratum and the shield, and there is a lack of a macro qualitative control method. The first step is to

(a)

(b)

(c)

**Fig. 5.3** **a** Photo of an event of collapse of the bridge deck of the short suspender fracture part of a bridge; **b** Slice scanning photo of an undamaged end of a short suspender; **c** Slice scanning photo of a damaged end of a short suspender; **d** Structural diagram of the bridge deck of a floating tied arch bridge; **e** Elevation of the scheme of the reconstruction of a bridge deck of a floating half-through tied arch bridge; **f** Detailed node drawing of the deck slab reconstruction scheme (Dimension unit: cm); **g** Plan of adding a limiting steel beam between the beam and the arch rib of the shortest suspender

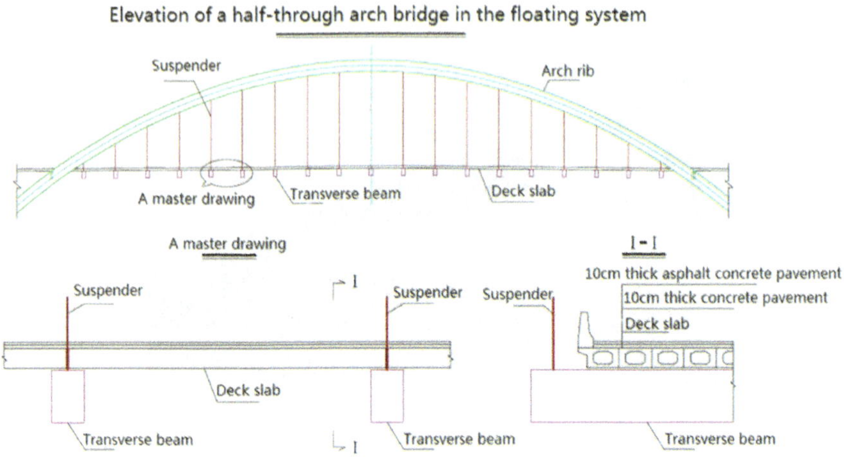


(d)

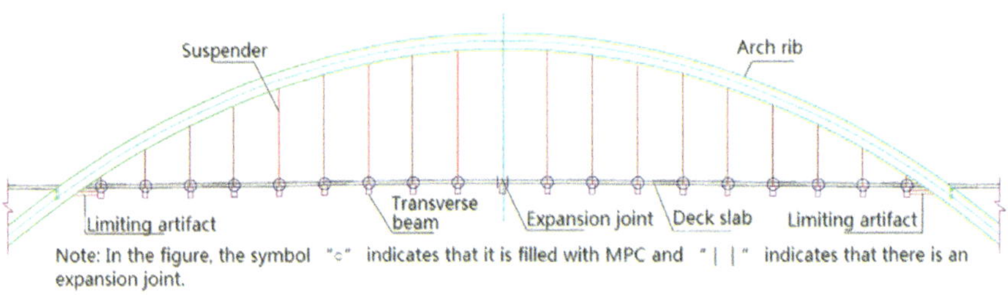


(e)

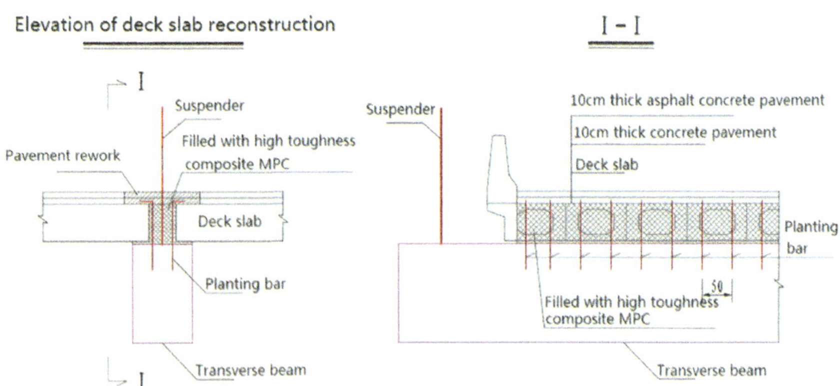


(f)

**Fig. 5.3** (continued)

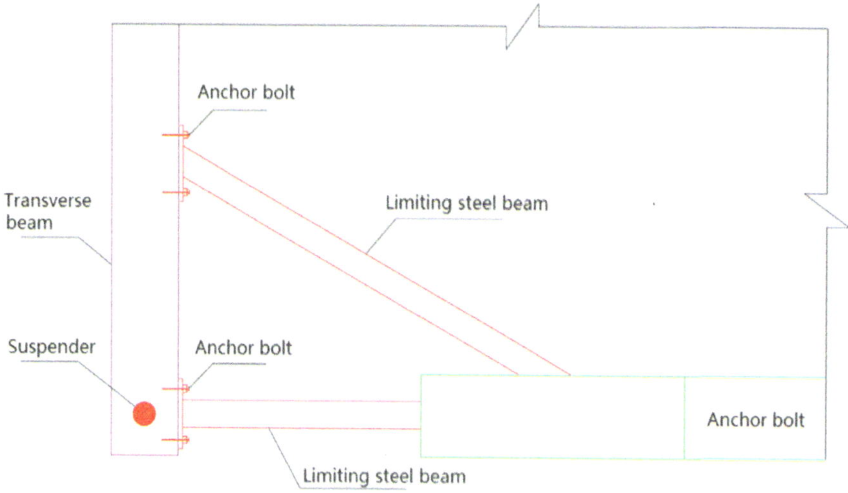


(g)

**Fig. 5.3** (continued)

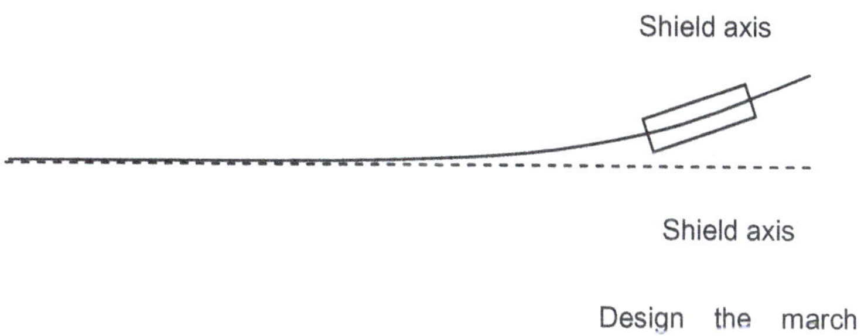


**Fig. 5.4** Diagram of the deviation of the shield axis march path

analyze the state of the position of the machine and the mechanical equilibrium to find the macroscopic qualitative control method (Fig. 5.5).

The thrust of the jack is simplified as $F_1$ at the top and $F_2$ at the bottom of the shield head; $F_{soil\,1}$ and $F_{soil\,2}$ represent the force of the soil reaction of the excavation face; $F_3$ is the elastic binding force of the machine body on the head. When the head floats up, it is prevented from floating up, and when the head is pressed down, it is prevented from pressing down. In addition, the shield is also affected by the buoyancy of the surrounding soil. The weight of $W_{shield} = 435.5\text{t}$, and the weight of the same volume of $W_{soil} = 577\text{t}$. Therefore, the buoyancy of the shield is 141.5t. The upper part of the formation is hard and the lower part is soft, which will delay the filling of the upper part of the gap without producing the resistance $F_4$ to prevent

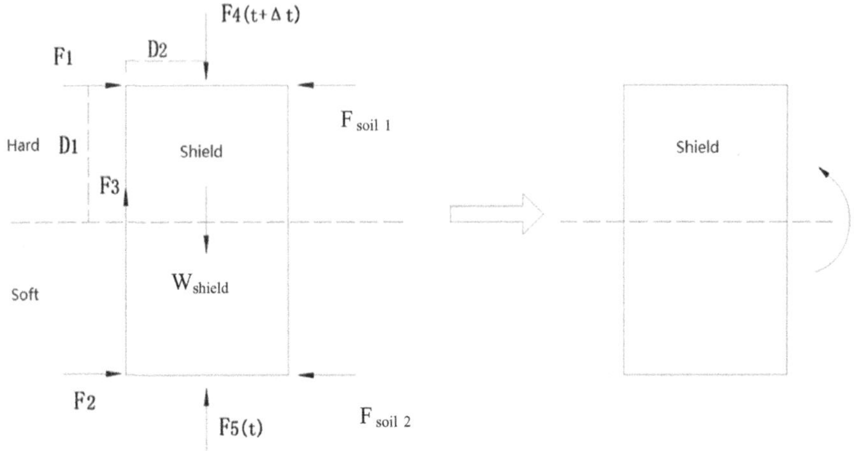


**Fig. 5.5** Diagram of the mechanical analysis of the position of the shield machine

the floating up; however, the lower part of the gap will be filled in time to generate the upward thrust $F_5$.

The conditions for a state of stable equilibrium of stand floating are as follows:

$$F_1 + F_2 = F_{soil\,1} + F_{soil\,2} \tag{5.1.1}$$

The above formula is easy to satisfy.

$$F_3 + F_4 + F_5 + W_{shield} = 0 \tag{5.1.2}$$

Because of the unfavorable deformation space and $W_{shield} - W_{soil} = -141.5t$, **only the simultaneous increase of F4 can satisfy** (5.1.2).

$$F_1D_1 - F_{\pm 1}D_1 + F_3D_2 - F_2D_1 + F_{\pm 2}D_1 = 0 \tag{5.1.3}$$

**Increase F1 or decrease F2 to meet** (5.1.3).

The second step is to formulate the adjustment plan and strictly control it (Fig. 5.6):

(1) Use formula (5.1.2) to synchronously inject the hard slurry with the weight greater than that of the upper-hard soil layer into the shield shell, eliminate the time difference $\Delta t \rightarrow 0$ in $F_4$ $(t + \Delta t)$, and **synchronously increase $F_4$** to basically control the position of the shield shell;
(2) At the same time, formula (5.1.3) is used to **increase $F_1$ or reduce $F_2$**, and the position of the shield shell is adjusted gradually to the axis of shield to design the traveling path.

**Fig. 5.6** Diagram of the relationship between shield tunneling construction and the surrounding water and soil equilibrium

**Problem 5.1.4**
With the expansion of the city, road construction goes through mountains and rivers. Most of the new tunnels belong to mountain tunnels, but about 200–400 m are located by the river. Shallow buried and concealed excavation is needed to pass through the soft flow plastic soil section. Shield construction is not economical. How should one design and construct to ensure the construction safety and the surrounding construction safety? For example, this is the case in Fig. 5.7.

Some design machines apply the shotcrete anchor support technology of NATM, which cannot control the deformation around the tunnel and the ground settlement at all during the construction, so the tunnel design should adopt the method of controlling the deformation around the tunnel and the ground settlement.

**Solution** In view of the shotcrete and anchor support technology for which the existing newly-built tunnel part needs shallow buried and concealed excavation to pass through the soft flow plastic soil stratum, the surrounding deformation and ground settlement of the tunnel cannot be controlled at all during the construction, and **the effective bearing structural layer should be formed in advance or in time** to meet the requirements of the control of the **effective bearing structural layer and the construction process**, so that it can **effectively mobilize and give full play to the self-bearing capacity of the underground engineering rock and soil mass (surrounding rock)**, improve the design and construction according to the coordinated control method of structural deformation, and pay attention so that the reasonable excavation and supporting method of tunneling is conducive to controlling the plastic deformation or loose deformation of the surrounding rock. **To ensure that the force is transferred according to the design, the unreasonable or even harmful transfer of the control force can avoid the sudden change or even collapse of the surrounding rock.**

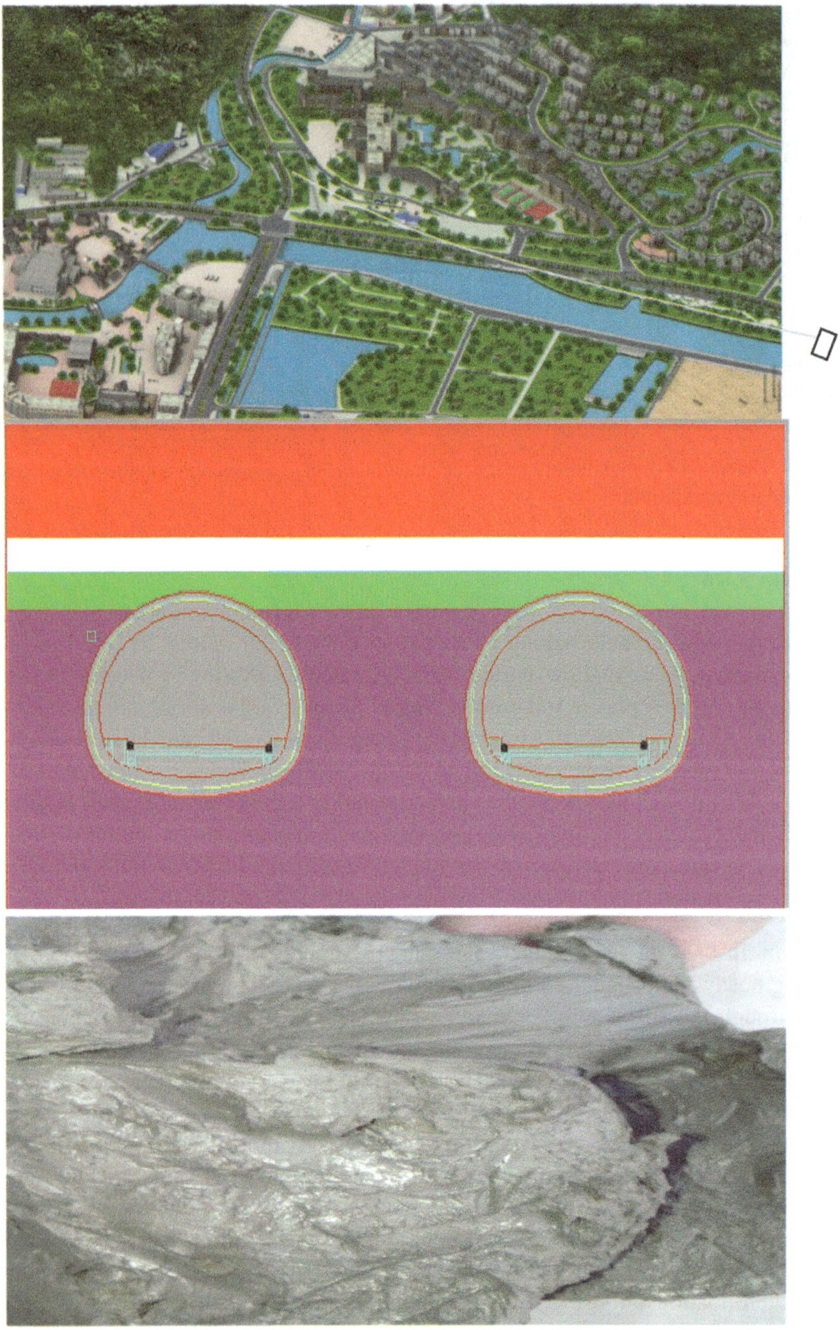

**Fig. 5.7** Part of the new tunnel needs shallow buried and concealed excavation to pass through the soft flow plastic soil stratum

When the tunnel adopts pre-supporting to keep the surrounding rock in its original state, there are:

$$P_1\cos\alpha_1 + P_2\cos\alpha_2 + T = W \quad (5.1.4)$$

In formula (5.1.4), $P_1$, $P_2$—mutual supporting force between surrounding rocks; W—gravity; T—supporting resistance (supporting resistance T is as small as possible). According to the formula (5.1.4), the size of the tunnel pre-supporting and initial supporting structure is estimated. For the soft flow plastic soil stratum, for the sake of safety, $P_1$ and $P_2$ are not considered, then (5.1.4) becomes

$$T = W \quad (5.1.5)$$

According to the formula (5.1.5), the size of the tunnel pre-supporting and initial supporting structure is estimated, and then the measurement results are revised.

The initial supports such as steel gusset peripheral supports, tunnel face grouting, improved excavation supporting method, steel arch frame and shotcrete bearing all loads, or horizontal jet grouting pile advanced support, forming the steel gusset peripheral supporting structure of the advance pipe shed limit can better meet the stability of the tunnel surroundings and even the tunnel face during the construction process.

(1) Support the surrounding of the steel gusset plate in advance of the pipe shed limit (Fig. 5.8): first of all, set a 108 mm long limiting steel pipe with an upward angle of 3–5° at an interval of 50, 5 cm outside the tunnel. The longitudinal lap length of each cycle is 4 m, and then the inner side of the limiting steel pipe is driven into the arc-shaped steel insert plate with the length of 8 m, the thickness of 10 mm and the width of 50 cm, and the longitudinal lap length of each cycle is 2 m. In this way, advance support is formed in the soft flow plastic soil layer outside the tunnel to control the deformation of the soft flow plastic soil layer outside the tunnel.
(2) Grouting on the tunnel face (Fig. 5.9):: first, insert a Ø20 mm plastic pipe with a grouting hole around 5 m in a quincunx shape of 50 * 50 cm in the tunnel face horizontally, with a longitudinal overlapping length of 1 m in each cycle. Then, according to the characteristics of the soft flow plastic soil stratum, the cement water glass double liquid slurry is prepared in the experiment, and the tunnel face is densely grouted. In this way, advance support is formed in the longitudinal soft flow plastic soil layer of the tunnel face to control the deformation of the longitudinal soft flow plastic soil layer of the tunnel face.
(3) Improve the method of excavation and support, the steel arch and shotcrete and other initial supports to bear all loads; under the condition that the deformation of the longitudinal soft flow plastic soil layer around the tunnel and the face is controlled in advance, the upper half cross-section is divided into ① and ② two pieces of longitudinal excavation of 1–2 m, and the support is closed in time. At this time, the initial support, such as the steel arch and shotcrete, bears

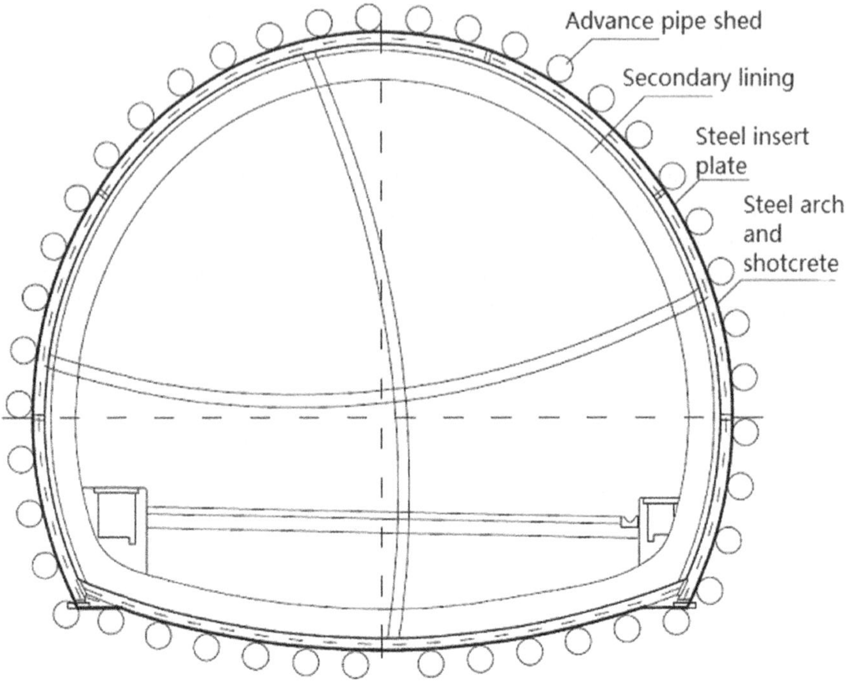


**Fig. 5.8** Structural diagram of the supports around the steel insert plate limited by the advance pipe shed

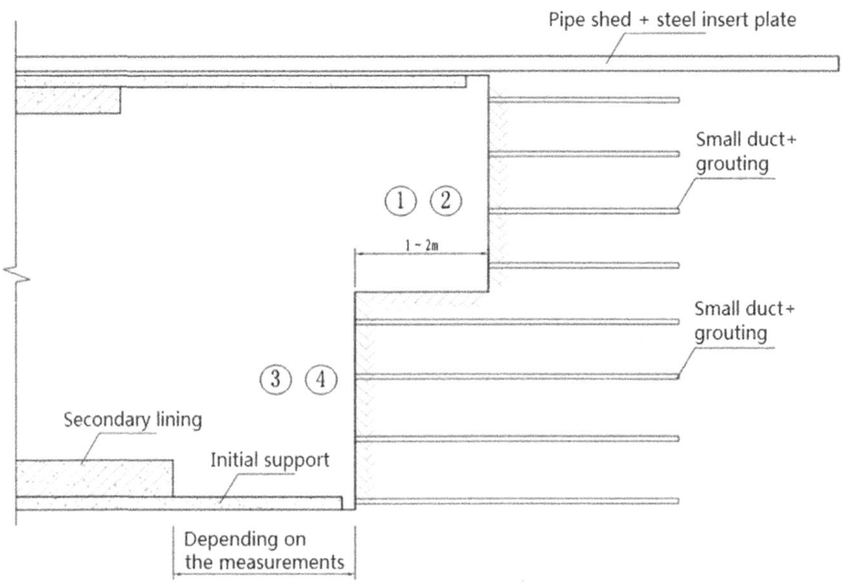


**Fig. 5.9** Tunnel face grouting and longitudinal excavation support

all the load, and then the lower half cross-section is divided into ③ and ④ two pieces of longitudinal excavation of 1–2 m, and the support is closed in time. At this time, the steel arch and shotcrete and other initial supports bear all of the loads. Each time, the upper half cross-section is divided into ① and ②, longitudinal excavation and support, and the lower half cross-section is divided into ③ and ④, two longitudinal excavations and circular supports, so as to control the deformation of the longitudinal soft flow plastic soil stratum around the tunnel and the face, which is conducive to the quality of the construction and the safety of the tunnel.

(4) Complete the secondary lining of the tunnel.

It is better to meet the stability of the surroundings and even the face of the tunnel during the construction process, effectively control the deformation and ground settlement around the tunnel, and ensure the safety of the tunnel construction and the normal use of the adjacent structures.

**Problem 5.1.5**

The problem of post-construction settlement of the existing multi-layer highway and railway soft soil subgrade is more prominent, especially the problem of post-construction settlement of a lane of soil-rock mixture filling of the bottom road at the side of the upper road column, which is reflected in the fact that the longitudinal bump and horizontal water accumulation of a lane of the bottom road can easily cause traffic accidents; however, for the integral subgrade such as lime soil subgrade and soil-rock mixture subgrade, there are very few phenomena of longitudinal bumps and horizontal water accumulation; the core problem of post-construction settlement of highway soft soil subgrade is to control the integrity of subgrade stress and deformation. How to solve the problem of integral control of the stress and deformation state of the new multi-layer highway and railway soft soil subgrade in the places where there are many coastal mountains and little land and abundant slag economically can avoid the repeated problems of the longitudinal bump and horizontal water accumulation of a lane of the bottom road, which would not only save construction funds but would also save maintenance funds!

**Solution** The settlement issue after the construction of the soft soil subgrade filled by the mixture of earth and stone in a lane of the existing upper highway and railway column and the bottom highway is prominent. That is to say, the problem of the integrity of the state of stress and deformation of the new subgrade filled with the mixture of soil and stone on the bottom layer is not effectively controlled. First, cement mixing piles or pre-stressed pipe piles are set up in the soft soil foundation, and then the integrity of the state of stress and deformation of the new subgrade filled with the mixture of soil and stone is controlled by the frame between the bottom layer of the subgrade and the top layer of the soft soil foundation. Generally, when the thickness of the soft soil layer is 15–20 m, a cement mixing pile is used for treatment; when the thickness of the soft soil layer is relatively large, the pre-stressed pipe pile is used for treatment; the cement mixing pile is recommended near the pier for treatment, so as to avoid traffic accidents caused by the longitudinal bump and horizontal water accumulation of a lane of the bottom road (Fig. 5.10).

**Problem 5.1.6**

There are many post-construction settlement phenomena of soft soil subgrade, especially the relatively poor characteristics of the soft soil subgrade near mountains and water, and the problem of lateral slip of the soft soil subgrade is more prominent. However, there are many methods of foundation treatment and analysis for the existing soft soil subgrade. Why are the successful cases not widely applicable? The two reasons that can easily be ignored are: (1) The main difference between the subgrade continuum and the dispersion: the dispersion can bear the pressure, but it cannot bear the tension or the moment basically; the continuum can bear pressure, tension and moment. **Controlling the integrity of the subgrade can provide the basement of a system of reaction equilibrium to resist the lateral slip of the subgrade; (2) The soft soil foundation of the soft soil subgrade has the characteristics of a large void ratio, high compressibility and so on. The phenomenon of lateral slip of soft soil subgrade can easily be caused. It is very important to control the reaction torque of the system of the reaction equilibrium of the soft soil foundation.**

In fact, the existing method of analysis and treatment of the lateral slip of the soft soil subgrade ignores the control of the system of the reaction equilibrium of the soft soil subgrade, especially when the characteristics of the soft soil subgrade near the mountains and water are relatively poor. It is not surprising that the problem of the lateral slip of the soft soil subgrade is more prominent (Fig. 5.11).

**Solution** In view of the two problems existing in the existing methods of analysis and treatment of soft soil subgrade, the traditional theory of the analysis of the soft soil subgrade ignores the control of the system of the reaction equilibrium of the soft soil subgrade. Only by simply controlling the integrity of the subgrade and providing the reaction torque can the lateral slip of the soft soil subgrade be resisted. On the basis of summarizing experience and dialectical thinking, the design of the soft soil subgrade is improved with the aim of controlling the integrity of the soft soil subgrade and providing the reaction torque.

The equation of the system of equilibrium and the system of the reaction equilibrium for the monolithic or bottom plate subgrade is as follows:

$$\mathrm{F_G} = \mathrm{F_M} \tag{5.1.6}$$

$$\mathrm{M_F} = \mathrm{F_M} * e \tag{5.1.7}$$

where, $\mathrm{F_G}$ is the subgrade weight; $\mathrm{F_M}$ is the subgrade reaction force; $\mathrm{M_F}$ is the subgrade reaction torque; e is the theoretical reaction torque of the system of the subgrade reaction equilibrium.

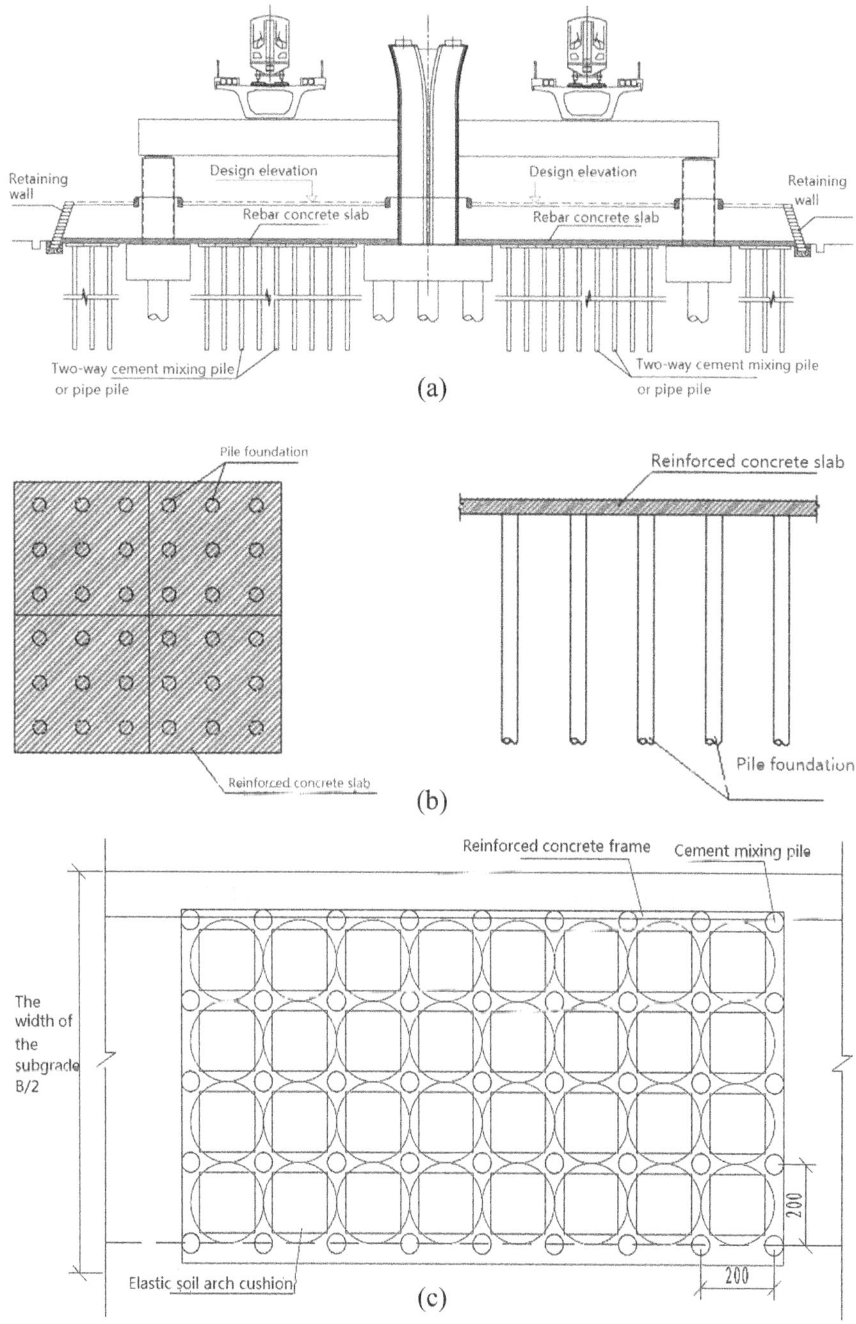


**Fig. 5.10** Soft soil subgrade treatment of a bottom layer highway. **a** Cross-section; **b** Plan layout and elevation; **c** Layout plan; **d** Vertical section layout; **e** Elevation of the elastic soil arch cushion

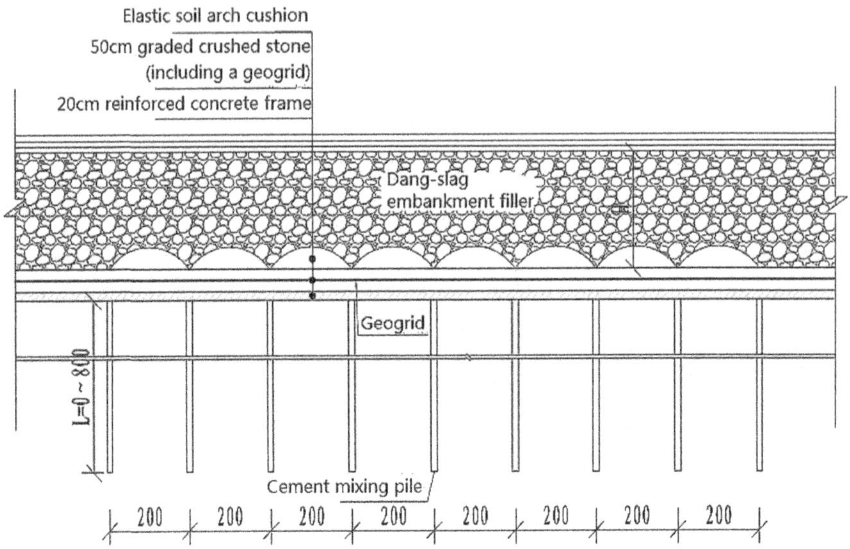


(d)

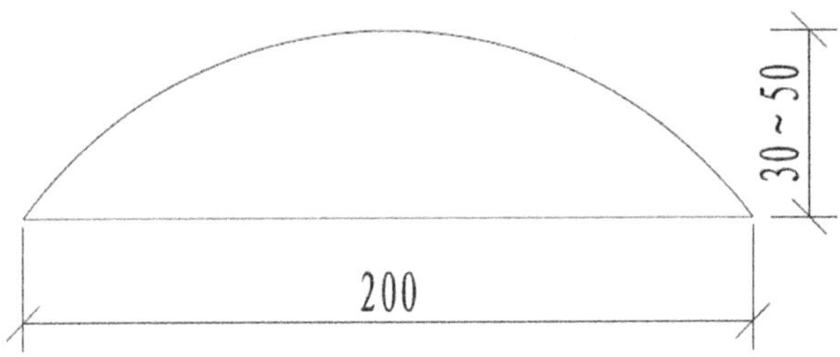


(e)

**Fig. 5.10** (continued)

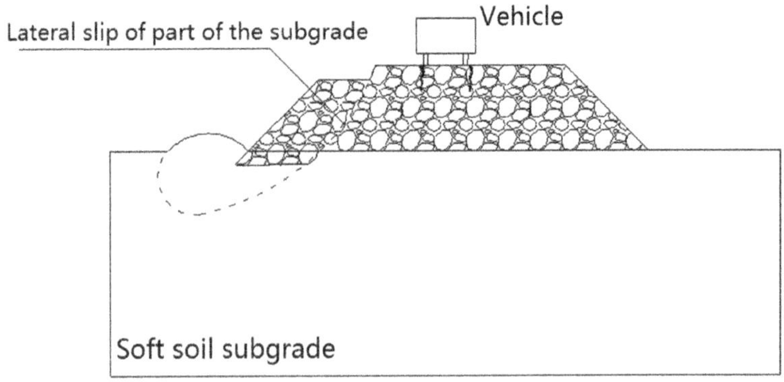


**Fig. 5.11** Lateral slip of the soft soil subgrade

**Therefore, formula** (5.1.6), (5.1.7) **and Fig. 5.12 show that according to the method of coordinated control of structural deformation, the method of designing the soft soil subgrade is improved.** In the design, the integrity of the state of subgrade stress and deformation and the lateral resistance pile are controlled. The key technology is adopted to control the bottom plate or frame and the lateral resistance pile of the subgrade post-construction settlement and construct the system of the reaction equilibrium of the soft soil subgrade, so as **to control the lateral slip of the soft soil subgrade.**

**Problem 5.1.7**

The construction method for eliminating the influence of welding residual stress of the steel truss arch bridge with full welded sections. Because of the large span and the heavy weight of the truss itself, the large-span steel truss arch bridge needs to be welded and assembled in multiple sections on site. At present, the cantilever assembly construction of this kind of bridge is usually carried out by the construction section truss first and then the weight of the construction section. When the cantilever length is large, the buckle cable is set to hold the cantilever section. The welding joint will continue to bear the load of the latter segment after the completion of the segment welding. Because the residual stress of the welding in the segment welding is not eliminated, the fatigue life will be affected.

For example, for a 336 m steel truss arch bridge with a main span, the construction method of "cable hoisting and cantilever splicing" is adopted for the installation of the main span arch rib, as shown in Fig. 5.13a. After cooling, the splicing weld of the segment begins to bear the vertical force and the bending moment generated by the next segment load and the pressure generated by the buckle cable. As the residual stress of the welding is not eliminated, it will have a harmful effect on the fatigue strength of the steel structure under dynamic load. Therefore, how to effectively eliminate the influence of the residual stress of the welding of the entire welded steel truss arch bridge is very important.

**Solution** Aiming at the full welded steel truss arch bridge, the construction method to eliminate the influence of the residual stress of the welding of the full welded segment assembled steel truss arch bridge is proposed. In this method, the temporary sling is used to fix the truss section to bear the vertical force of each section, and a load-bearing cable is added to fix each temporary sling. As shown in Fig. 5.13b, c, d, i.e. lifting the truss section into place with a lifting sling, welding after temporary bolt anchoring, and then installing the temporary sling fixed to the load-bearing cable. The temporary sling fully bears the weight of the truss. After the temporary sling is set, the control of the deformation coordination of the truss segment can be constrained. The spliced truss can no longer bear the weight of the later segment, so the spliced weld is no longer stressed, and the residual stress of the welding caused by the external load can be gradually eliminated.

Adding a temporary sling to the load-bearing cable will change the force of the cable of the original temporary sling, so it is necessary to adjust the force of the cable

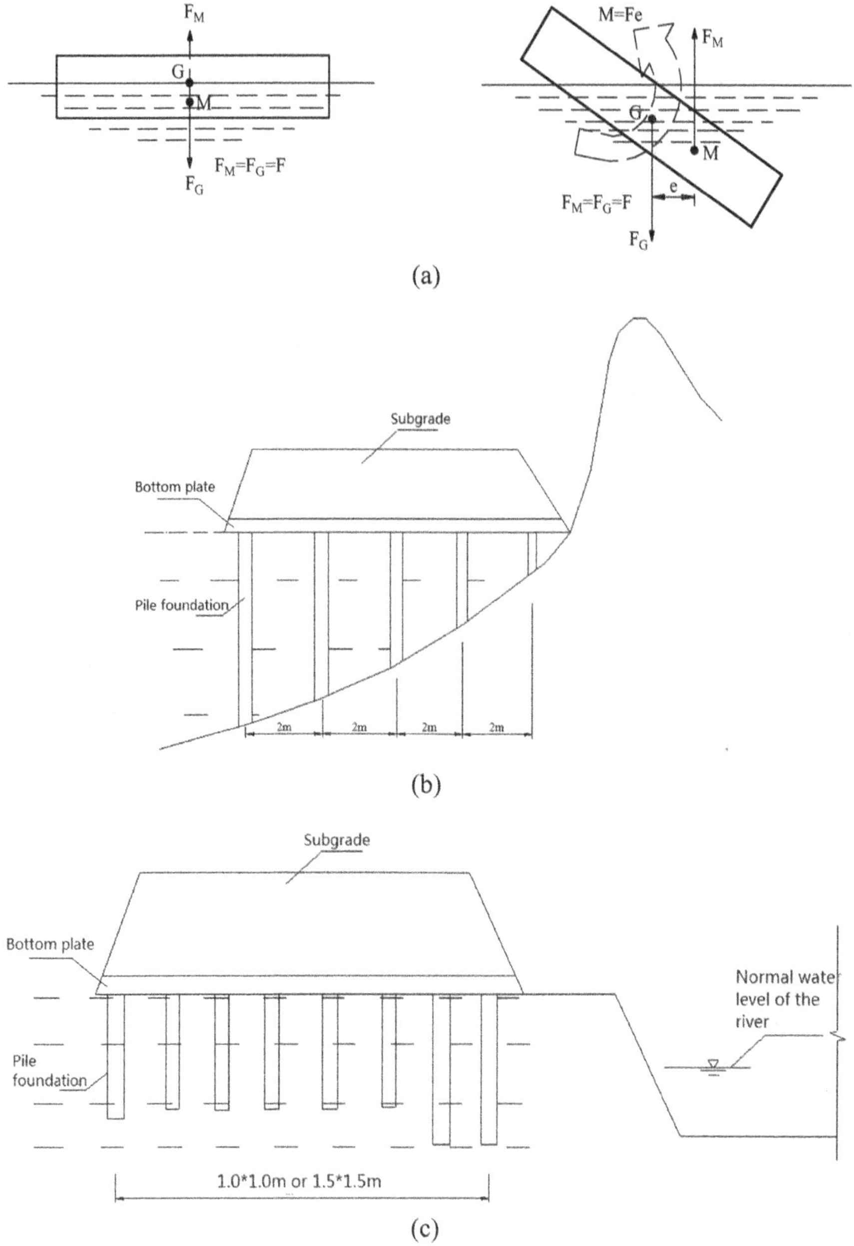


**Fig. 5.12** **a** Diagram of the stress analysis of the reaction torque treatment measures for controlling the system of the equilibrium of the soft soil foundation reaction; **b** and **c** Structural drawing of the bottom plate or frame and the lateral resistance piles that control the lateral slip of the soft soil subgrade on the mountain side by adopting this method

**Fig. 5.13** **a** Cantilever assembly construction of a bridge; **b** Vertical layout of hoisting; **c** Hoisting layout plan; **d** Diagram of the sequence of the hoisting construction

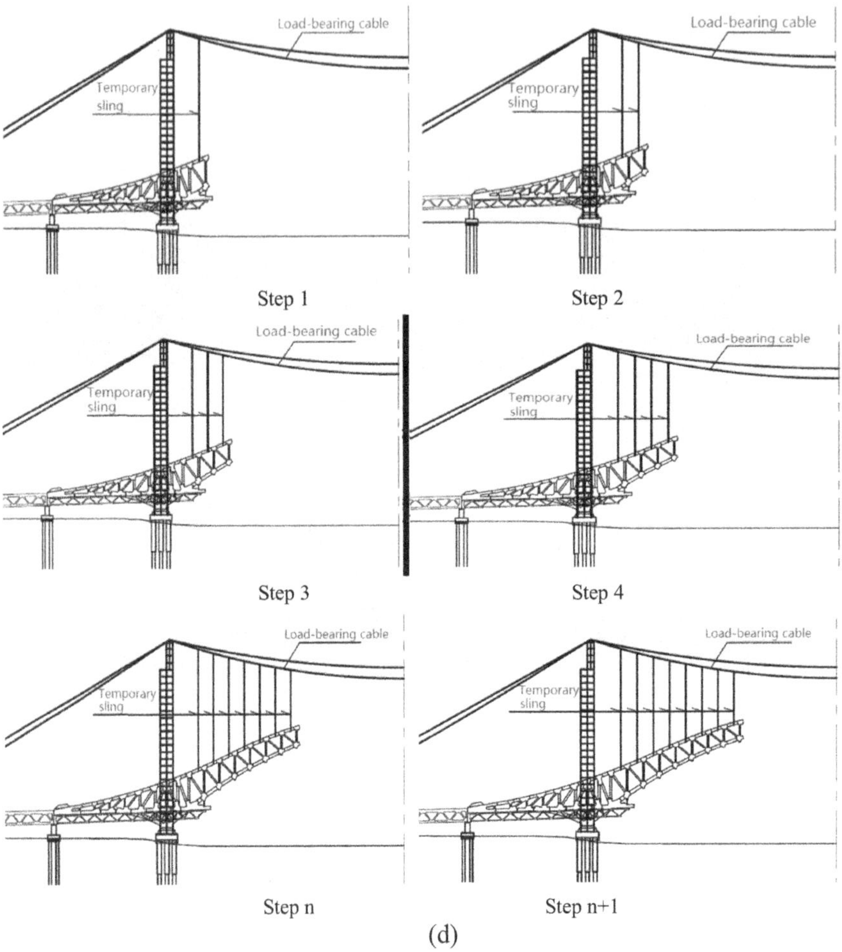


**Fig. 5.13** (continued)

of the used temporary sling in the process. In addition, due to the large load on the load-bearing cable, there should be reliable back anchors at both ends of the cable.

Construction technology of the load-bearing cable: (1) Steel wire rope, high-strength steel wire, etc. can be used as materials, and anti-corrosion treatment. (2) The load-bearing cable is set on the cable tower, and both ends of the cable are anchored on the temporary anchorage.

Construction technology of the sling: (1) Steel wire rope, high-strength steel wire, steel strand, etc. can be used as materials, and anti-corrosion treatment can be carried out. (2) One end of the sling is fixed onto the cable and the other end is fixed onto the truss. One end of the sling is the tension end, which is used to adjust the force of the cable.

**Problem 5.1.8**

At present, the research on steel bridge deck pavements in China shows that there are two-layer SMA Pavements in the early stages, but there are serious diseases such as displacement, cracking, rutting and so on, and they are not used now. Domestic professionals have conducted in-depth research on the failure of the double-layer SMA mixture in being successfully applied to the steel bridge decks in China, and this research found the root cause of the problems, that is, that the interface structure material does not meet the pavement requirements.

Specifically, the interface material of SMA paved steel bridge decks is composite modified high viscosity asphalt, which is coated on the steel plate to make the SMA and steel plate bond together. However, the adhesive force of the modified high viscosity asphalt stiffening agent to the steel plate decreases rapidly with the increase in the steel plate temperature. When the temperature rises to a certain extent, the adhesive force cannot meet the bonding requirements, and then the SMA layer will produce a sliding displacement on the steel plate, it will push convexly and generate cracks.

After that, American double-layer epoxy asphalt and British pouring asphalt pavement technology were introduced, but there were also many problems, which were not solved completely;

American epoxy asphalt pavement technology has the advantages of high pavement strength, good integrity, resistance to high temperature deformation, resistance to low temperature cracks and strong resistance to corrosion. However, its disadvantages are a high cost, difficult construction, difficulty in repairing after damage, and strict requirements on construction environment. Construction control that is not strict is the main cause of pavement damage, and there is no repair method for epoxy asphalt damage.

The cast asphalt pavement technology in Britain has low porosity, strong waterproof capability and good anti-aging properties. It has the advantages of strong resistance to cracking and good adhesion to the steel plate, but its disadvantages lie in poor high-temperature stability, easy formation of ruts, the need for specific construction equipment and complex organization of construction.

ERS cold mix resin asphalt steel deck pavement technology and steel fiber reinforced concrete steel deck pavement technology solve the above problems, but there are also inconveniences in construction and maintenance.

To sum up, it is necessary to develop a steel bridge deck pavement material and its technology to better solve the above problems, which cannot only meet the special situation of high temperatures and heavy loads in China, but that can also avoid the shortcomings of the above four technologies.

At present, the steel deck pavement technology of PVA ductile fiber-reinforced concrete has solved the above problems. The expansion coefficient of PVA ductile fiber-reinforced concrete layer and steel deck is consistent, and the two are relatively firm. Especially in the case of large temperature differences and vehicle loads, PVA ductile fiber-reinforced concrete layer and steel decks can also be well bonded together, with strong anti-sliding ability. It further prevents cracks, ruts, bumps and other diseases of the bridge deck pavement. Construction and maintenance are also convenient, but the structural reinforcement mesh is set in the middle or on the upper position, and this is not conducive to the control of structural deformation coordination.

Therefore, the structural reinforcement mesh is set in the middle lower position of the PVA ductile fiber-reinforced concrete layer, so that the steel deck pavement can not only control the insufficient rigidity of the steel deck plate, but it can also control the decreasing vertical rigidity of the pavement itself to change the asphalt concrete surface, adapt it to the deformation coordination of the steel deck and the action of the vehicle load, and render it conform to the method of control of structural deformation coordination, which is a better technology of steel deck pavement.

**Solution** In view of the shortcomings and defects of the steel deck pavement technology mentioned in the above technical background, a method of control of the deformation coordination of steel-PVA ductile fiber-reinforced concrete composite deck pavement is provided.

This kind of structure has the advantages of low building height, light weight and high strength, good bonding performance between composite layers, uniform cooperative stress of each composite layer, deformation coordination that is easy to control, good durability, good resistance to fatigue and a small impact effect from vehicles.

The steel deck pavement technology of PVA ductile fiber-reinforced concrete is adopted, in which the structural reinforcement mesh is set at the position of the middle lower part of the PVA ductile fiber-reinforced concrete layer. In this way, the steel deck pavement can not only control the insufficient rigidity of the deck steel plate, but it can also control the decreasing of the vertical rigidity of the pavement itself. It can adapt to the deformation coordination of the steel deck and the driving load, and conform to the method of control of the structural deformation coordination.

The composite deck structure includes a steel deck layer, a PVA ductile fiber-reinforced concrete layer poured over the steel deck layer from bottom to top. The steel bridge deck layer is provided with a shear structure, a tensile structure and a neoprene cushion block. The shear structure is a consolidated shear bolt on the steel bridge deck. The tensile structure is a reinforcing mesh formed by the superposition of a longitudinal tensile reinforcing steel distribution layer and a transverse tensile reinforcing steel distribution layer. The shear bolt, the reinforcing mesh and the neoprene cushion block are embedded in a PVA ductile fiber-reinforced concrete layer.

The shear studs are arranged in a matrix interval along the longitudinal and transverse direction of the steel bridge deck. The longitudinal and transverse spacing between the adjacent shear studs should conform to the specifications and design requirements, and can meet the shear requirements of the steel-PVA ductile fiber-reinforced concrete interface. The shear bolt is prefabricated in the factory according to the design requirements, and the length of the shear bolt is preferably from 8 mm to 12 mm. The contact area between the shear bolt and the steel bridge deck layer within this range is larger than that between the traditional shear bolt and the steel bridge deck layer. Therefore, the stiffness of the shear studs is greater than that of traditional shear studs, which can improve the shear efficiency of the shear studs and further improve the combined stress between steel-PVA ductile fiber-reinforced concrete.

The tensile structure of the reinforcing mesh is formed by the superpositioning of the longitudinal tensile reinforcement distribution layer and the transverse tensile reinforcement distribution layer. The longitudinal reinforcement distribution layer in the tensile structure of the reinforcing mesh is laid on the neoprene cushion block. The transverse reinforcement distribution layer in the tensile structure of the reinforcing mesh is laid on the upper part of the longitudinal reinforcement distribution layer, the longitudinal reinforcement distribution layer and the transverse reinforcement distribution layer are connected by binding or spot welding, the tensile structure of the reinforcing mesh is set at the middle lower position of the PVA ductile fiber-reinforced concrete layer, and is located in the upper part of the shear bolt. This setting can make the steel deck pavement not only control the insufficient rigidity of the steel deck, but it can also control the decrease in the vertical rigidity of the pavement itself. It can adapt to the deformation coordination of the steel deck and the driving load, and conform to the method of control of the structural deformation coordination. The transverse tensile steel bars in the distribution layer of the transverse tensile steel bars of the tensile structure of the steel mesh and the lower shear studs are consolidated by spot welding, and the shear studs and the steel mesh are embedded in the PVA ductile fiber-reinforced concrete layer.

The neoprene cushion block is arranged to support the longitudinal tensile reinforcement, to facilitate the laying of the longitudinal tensile reinforcement, and to facilitate the binding or spot welding connection of the longitudinal tensile reinforcement and the transverse tensile reinforcement.

The PVA fiber toughness concrete layer is poured above the steel bridge deck, and the PVA fiber toughness concrete layer is poured from the PVA fiber toughness concrete. The PVA fiber reinforced concrete is made up of PVA fiber, cement, fine sand, flyash, water and a water reducing agent in a certain proportion, and the optimal mix is shown in Table 5.3. The PVA ductile fiber-reinforced concrete has excellent mechanical properties, its compressive strength is 35–60 MPa, its tensile strain capacity is 300–500 times of that of ordinary concrete and fiber-reinforced concrete, its bending strength can reach 3–5 times that of its direct tensile strength, and its durability coefficient is 5 times that of ordinary concrete. The deformation coordination between PVA fiber-reinforced concrete and reinforcement is consistent, and the crack width can be controlled to around 60 $\mu$m, which makes the permeability coefficient of PVA fiber-reinforced concrete layer in the cracked state almost the same as that of plain concrete without cracks, effectively improving the durability and impermeability of the components.

**Table 5.3** ECC mix proportion (kg/m$^3$)

| Cement | Fine sand | Flyash | Water | Water reducer | PVA fiber |
|---|---|---|---|---|---|
| 577 | 462 | 692 | 323 | 9 | 26 |

A bonding layer is laid on the upper part of the PVA ductile fiber-reinforced concrete layer, which is epoxy asphalt adhesive. The epoxy asphalt adhesive has the characteristics of high bonding strength, high toughness and resistance to fatigue. The epoxy asphalt adhesive is firmly bonded with the PVA ductile fiber-reinforced concrete layer and asphalt concrete surface layer to form an integral layer after curing at room temperature, effectively improving the integrity of the combined bridge deck.

The asphalt concrete surface layer is laid on the bonding layer. The asphalt concrete surface layer can be all kinds of asphalt concrete bridge deck pavement or bridge pavement of other materials. According to the needs of the engineering technical design, the appropriate asphalt concrete surface pavement layer can be selected flexibly. The construction is convenient, the process is simple, and has good operability and is economic.

Compared with the existing technology, the advantages of deformation coordination control structure of the steel-PVA ductile fiber-reinforced concrete composite deck pavement are as follows (Fig. 5.14):

(1) The PVA ductile fiber-reinforced concrete layer in this technical proposal is suitable for steel bridge decks. It has excellent mechanical properties, high compressive strength, bending strength and tensile strain capacity. Its durability coefficient is also higher than that of ordinary concrete, and the expansion coefficient of PVA ductile fiber-reinforced concrete is consistent with that of steel bridge decks, and the two are relatively bonded. It is firm, especially in the case of large temperature differences and vehicle loads, PVA ductile fiber-reinforced concrete layers and steel bridge decks can also be bonded together well, with strong anti-slip ability, so as to further prevent the cracks, ruts, bumps and other problems of the bridge deck pavement.
(2) The application of PVA ductile fiber-reinforced concrete on steel bridge decks can effectively reduce the thickness of the concrete layer, reduce the weight of bridge deck system, and significantly improve the durability of the composite bridge deck.
(3) By controlling the height of the shear bolt set on the steel deck and the consolidation area of the steel deck, the design height of the shear structure between the steel deck and the PVA fiber-reinforced plastic concrete layer can also be adjusted flexibly, so as to reduce the design thickness of the PVA fiber-reinforced plastic concrete layer and the weight of the composite deck structure itself. At the same time, by adjusting the effective area of shear studs fixed on the steel deck, the shear stiffness of the composite deck can be further increased, so that the stress of the composite deck is more reasonable.
(4) The design of the shear bolt of the composite bridge deck structure can further reduce the relative slip between the bridge deck pavement and the steel bridge deck, it can reduce the convex hull of the bridge deck pavement, reduce the effect of the impact of the vehicle load, and further improve the durability of the bridge deck pavement.
(5) The shear stud used in the shear structure is prefabricated in the factory according to the design requirements, and the setting of the shear structure does

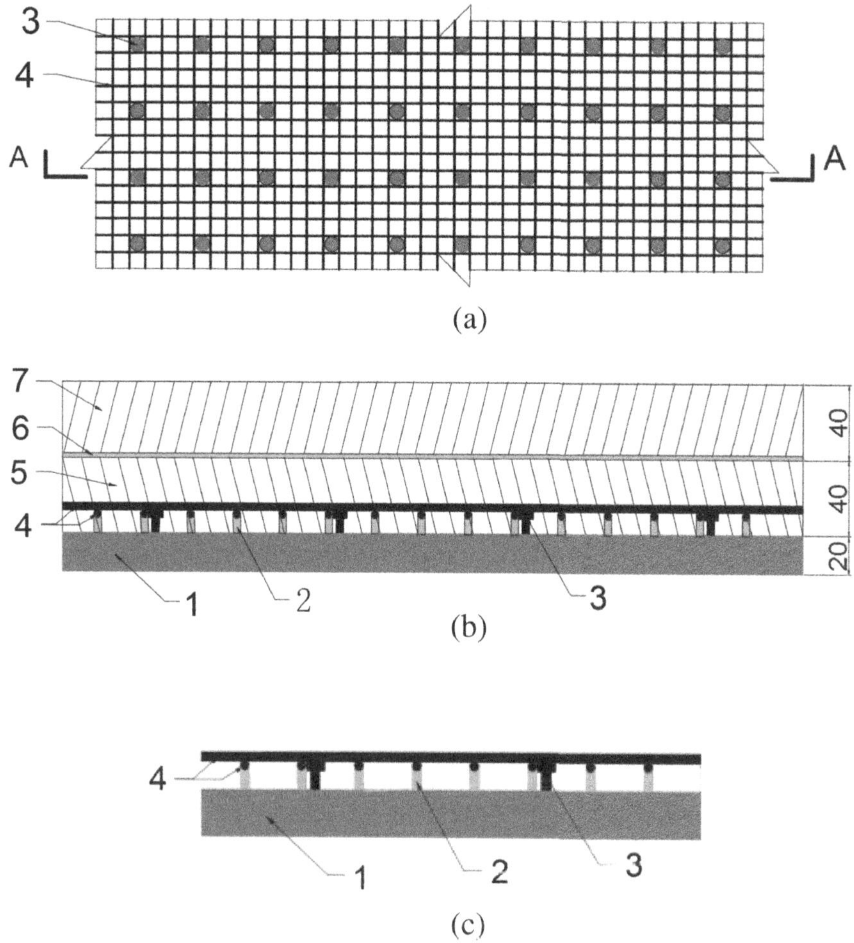


**Fig. 5.14** **a** Perspective view of the composite deck plane pavement; **b** Structural diagram of the composite deck; **c** Local structure of the composite deck (PVA concrete not shown)

not need to use complex construction technology and high input construction equipment. The shear bolt can be consolidated with the transverse tensile steel bar arranged on the upper part of the shear bolt by ordinary welding technology. The investment in the equipment is small, simple and easy to operate, and the quality of the labor and the requirements of the process are low.

(6) Setting the tensile structure of the steel mesh in the middle lower position of the PVA ductile fiber-reinforced concrete layer can make the steel deck pavement not only control the insufficient rigidity of the deck steel plate, but it can also control the decrease in the vertical rigidity of the pavement itself of the transition asphalt concrete surface, adapt to the deformation coordination

of the steel deck and the driving load, and conform to the control method of the structural deformation coordination.

(7) The bonding layer is epoxy asphalt adhesive, which has the characteristics of high bonding strength, high toughness and resistance to fatigue. After curing at room temperature, the epoxy asphalt adhesive is firmly bonded with the PVA fiber reinforced concrete layer and the asphalt concrete surface layer to form an integral layer, which effectively improves the integrity of the combined bridge deck.

(8) The above-mentioned epoxy asphalt adhesive can be solidified at a normal temperature without special requirements for the construction environment and process, so no special construction equipment is needed during construction, and the construction can be carried out according to the laying method of a common bonding layer.

(9) The adopted asphalt concrete surface course has no special requirements for the construction environment and technology. It can be the bridge pavement of various types of asphalt concrete bridge deck pavements or other materials. It can flexibly select the appropriate asphalt concrete surface course according to the needs of the engineering technical design, which is convenient for construction, simple in technology, and has good operability and is economic.

(10) To sum up, the deformation coordination control structure of steel-PVA ductile fiber-reinforced concrete composite deck pavement has the advantages of small building height, light weight and high strength, large rigidity, good bonding performance between composite layers, uniform cooperative stress of each composite layer, easy control of deformation coordination, good durability, good resistance to fatigue, a small effect of the impact of vehicles, etc., which has great practical value and good economic efficiency. It is especially suitable for steel deck pavements of long-span bridges.

**Problem 5.1.9**

In recent years, with the rapid development of highway bridge construction in China, the number of fires caused by inflammables under highway bridges, especially under the bridges at the junction of urban and rural areas, is increasing due to many influencing factors, and this is difficult to control. Fire not only causes serious economic losses, but also the mechanical properties of reinforced concrete structures such as strength and elastic modulus will decrease under the physical and chemical effects of fire, the bonding strength between reinforcement and concrete will decrease correspondingly, and the bearing capacity of the bridge section will also decrease, which will cause the hidden danger of bridge safety and will affect the safety of highway operations. Therefore, it is necessary to make a correct judgment quickly on the degree of damage of the bridge structure, then adopt scientific and reasonable repair and reinforcement methods, resume the operation as soon as possible, and reduce the loss caused by fire to the minimum, which is an urgent problem to be solved at present. The "Experimental Research and Practical Application of the Steel Plate and

Concrete Composite Reinforcement Method" (Shi Xiongwei 2012) has disclosed a common reinforcement method of bridges over fire in the existing technology, which mainly includes: composite reinforcement for beams in severe fire areas, and bonding steel plate reinforcement for beams in the general areas. The steel plate is fixed onto the hollow slab by the reinforcement, and the dense concrete is used for filling between the hollow slab and the steel plate. This method can play a role of reinforcement in theory, but the applicant found that in the actual process of use, on the one hand, because its steel plate only goes over the fire bottom plate and part of the web, so the load will gradually focus on the web not covered by the steel plate in the subsequent process of use, which will cause hidden danger to the structural safety of this part of the area; on the other hand, its steel plate is planted with reinforcement which will weaken the flexural rigidity of the beam and affect its bearing capacity. At the same time, when the dense concrete is used for filling, the bond performance between the dense concrete and the original concrete and steel plate is insufficient, and so can easily fall off. The above construction method needs to lay the support, which not only takes time and affects the speed of the operation, but it also cannot be used in the areas where it is difficult to erect the support under the bridge with great demand of traffic. Therefore, it is necessary to provide a better reinforcement method for bridges damaged by fire.

**Solution** The method for strengthening a bridge damaged by fire is aimed at one of the hollow slab beams or T-beams or the small box girders (Fig. 5.15). The main purpose of the method for strengthening is to improve the vertical stiffness and integrity of the strengthened bridge, reduce the difference between the load distribution ratio and the design, and improve the bearing capacity of the bridge.

The method for strengthening the hollow slab beam is as follows:

1.1 Remove the bridge deck pavement. Chisel out the hinge joint of the beam and slab, and use planting reinforcement to strengthen it;
1.2 Remove the loose and peeling concrete under the hollow slab beam damaged by fire by means of manual chiseling. The concrete on the surface of the beam and slab should be cleaned using a high pressure water gun, and the rusted reinforcement should be de-rusted and cleaned, and then the reinforcement should be treated with rust prevention;
1.3 The lifting equipment should be erected at both ends of the beam slab hinge joint between the two cover beams. The lifting rope should be used to lift the H-shaped steel beam or double jointed channel steel in place through the hinge joint, and then the connecting screw should be connected. Two strip baffles should be pasted on the upper flange of the H-shaped steel beam or double-jointed channel steel to form a cavity between the upper flange, the strip baffle and the bottom surface of the beam slab with a hinge joint; at the same time, fix the steel plate at the hinge joint above the cover beam. The steel plate plays the role of sealing the hinge joint above the cover beam, which can be set separately or extended by the upper flange of the steel beam;

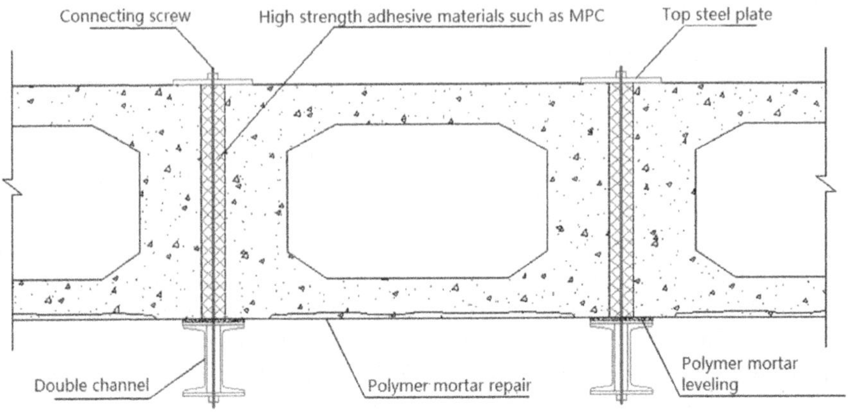


(a)

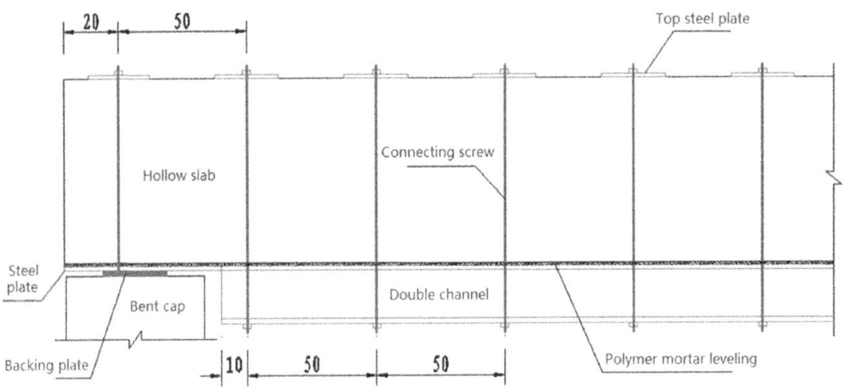


(b)

**Fig. 5.15** **a** Cross-section of the reinforcement scheme of a hollow slab damaged by fire; **b** Elevation of the reinforcement scheme of a hollow slab damaged by fire; **c** Cross-section of the reinforcement scheme of a T-beam damaged by fire; **d** Cross-section of the reinforcement scheme of a small box girder damaged by fire; **e** Elevation of the reinforcement scheme of a T-beam (or small box girder) damaged by fire

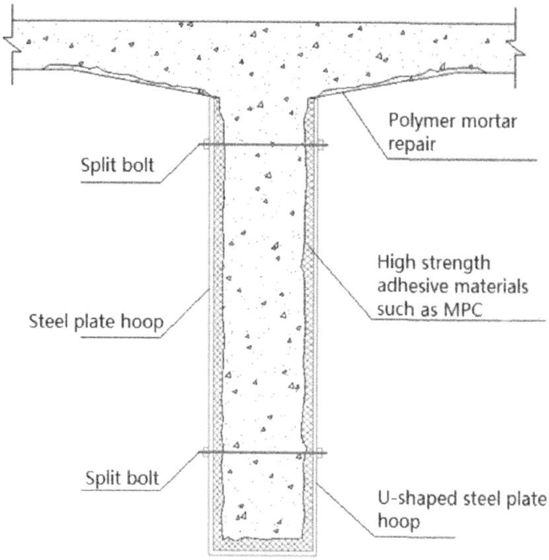


(c)

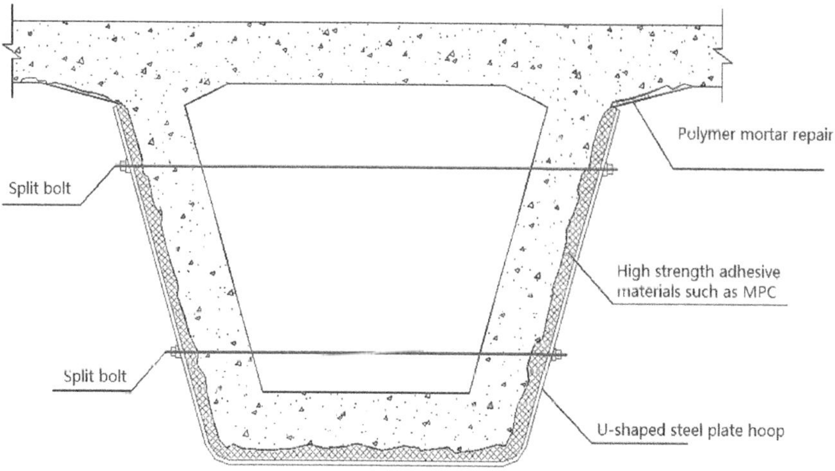


(d)

**Fig. 5.15** (continued)

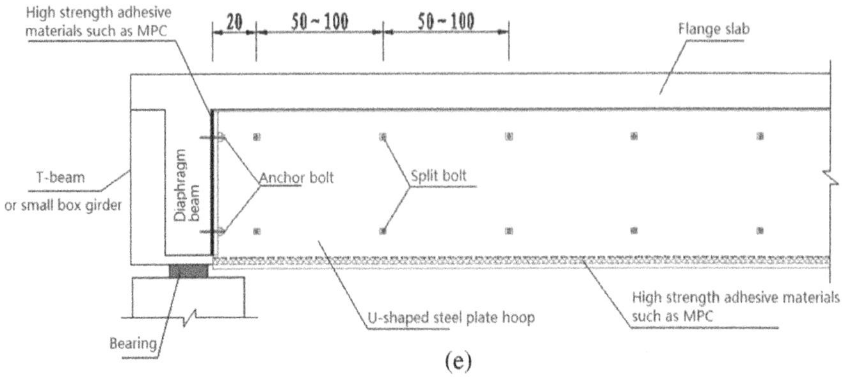

(e)

**Fig. 5.15** (continued)

1.4 Pour the high-strength bonding material into the hinge joint of the beam and slab, make the high-strength bonding material fill the cavity and hinge joint, and glue the beam and the steel beam into a whole, so as to play the role of common stress and deformation;
1.5 Other chiseled concrete parts should be repaired with polymer repair mortar for peeling.

The method for strengthening the T-beam is as follows:

2.1 The loose peeling concrete under the T-beam damaged by fire should be chiseled manually. The concrete on the surface of the beam and slab should be cleaned using a high-pressure water gun, the rusted reinforcement should be de-rusted and cleaned, and then the reinforcement should all be treated for rust prevention;
2.2 Drill holes at the wet joints of the bridge deck, erect hoisting equipment above the T-beam, hoist the U-shaped steel hoop in place through the holes drilled at the wet joints, make the U-shaped steel hoop cover the whole bottom plate and web of the T-beam, and then connect the split screw to fix the U-shaped steel hoop on the T-beam;
2.3 Fill the T-beam and the U-shaped steel hoop with high-strength bonding material, and bond the T-beam and U-shaped steel hoop into a whole, playing a role of common stress and deformation;
2.4 Other chiseled concrete parts should be repaired with polymer repair mortar to the peeling part.

The method for strengthening the small box girder is as follows:

3.1 The loose and peeling concrete under the small box girder damaged by fire should be chiseled manually, the concrete on the surface of the beam and slab should be cleaned using a high-pressure water gun, the rusted reinforcement shall be de-rusted and cleaned, and then the reinforcement should be treated for rust prevention;

3.2 Drill holes at the wet joints of the bridge deck, erect hoisting equipment above the small box girder, hoist the U-shaped steel hoop in place through the holes drilled at the wet joints, make the U-shaped steel hoop cover the whole bottom plate and web of the small box girder, and then connect the split screw to fix the U-shaped steel hoop on the small box girder;
3.3 Fill the small box girder and the U-shaped steel hoop with high-strength bonding material, and bond the small box girder and U-shaped steel hoop into a whole to play a role of common stress and deformation;
3.4 Other chiseled concrete parts should be repaired with polymer repair mortar.

The U-shaped steel plate hoop refers to a structure formed by welding steel plates, which is generally U-shaped or concave.

In the above reinforcement process, first of all, it is necessary to chisel out the loose peeling concrete under the hollow slab beam damaged by fire. If it is not chiseled and repaired directly, this part could easily fall off in the subsequent process of use and affect the integrity. Furthermore, the steel plate/steel beam should be used to reinforce the hinge joint of the beam and slab. The bottom plate and web of the T-beam or small box girder, and high-strength bonding materials such as MPC should be poured at the same time, so that the steel plate/steel beam, bonding materials and original beam body can form a whole, suffer deformation together, and prevent excessive load from gathering at a vulnerable part, resulting in a potential safety hazard. For the construction of the hollow slab beam, the method without support is adopted, that is, by erecting hoisting equipment directly at both ends, lifting the steel beam to the bottom of the beam slab, and pouring bonding materials into it through the hinge joint. This method can greatly improve the convenience of construction in the area where there is a demand for traffic under the bridge and the support is not suitable.

The high-strength bonding material must meet the following requirements: (1) The bonding strength with concrete must not be less than 2.5 MPa; (2) The quantity of the elastic model must be more than 1000 MPa; (3) After mixing, the setting time must be more than half an hour; (4) The bonding strength within 6 h must be more than half of the standard strength.

The following describes the basis for selecting the above indicators of high-strength bonding materials. The role of bonding materials is divided into two aspects: One is to improve the overall structural strength between different hollow slab beams, T-beams or small box girders; the other is to bond the double channel steel or H-shaped steel beams to the bottom of the hollow slab beams and the side of the T-beams or the small box girder cladding. (1) Generally, the anti-crack strength of concrete is 2 Mpa, so the bonding strength of concrete should not be less than 2 Mpa, which is optimized to be not less than 2.5 MPa through testing, with a safety factor reserved. When the bonding materials meet the requirements, the bond between the different beams and slabs will not be damaged even if the load is too high, and the opposite damage will occur in the beam slab concrete. Therefore, it greatly improves the integrity between beams and slabs. (2) In addition, through testing, it was found that when the quantity of elastic model is greater than 1000 MPa, the bending deformation of the steel beam can be coordinated, otherwise, the steel beam with a quantity of elastic model that is too small may not be pasted on the beam plate, the T-beam or the small box girder, resulting in partial or complete separation of the two, thus unable to play the role of improving the rigidity of the beam body. (3) Considering the needs of the construction process, the setting time after mixing cannot easily be too short, otherwise the best bond performance cannot be achieved in the pouring process. (4) In consideration of the efficiency of construction, the bonding strength within 6 h should be more than half of the standard strength, otherwise the required time will be too long, which will affect the operationing and opening of the bridge.

The high-strength bonding material is epoxy mortar or MPC composite material. The two kinds of materials can be applied to the reinforcement of bridges, and the construction is suitable.

The double channel steel adopts two hot-rolled channel steel and upper and lower steel plates welded as a whole, and is fixed under the beam plate hinge joint between the two cover beams through the connecting screw, while the upper steel plate of the double channel steel extends to cover the hinge joint above the cover beam; The double channel steel or H-shaped steel beam meets the following conditions: (1) The width of the upper flange bonded to the beam bottom plate should not be less than 150 mm, the height should not be less than 150 mm, and the thickness of the plate should not be less than 6 mm; (2) Q345 or Q235 steel should be used; (3) Weathering steel, carbon steel and low alloy high strength steel should be used; (4) The upper surface of the upper flange of the steel beam should be shot blasted Sa2.5 in the factory, if rust is found on site, the shot should be manually blasted to St3.0. The function of shot blasting is to increase the bonding strength of the bonding materials.

When carbon steel or low-alloy high-strength steel is used for the double channel steel or H-shaped steel beam, except for the upper flange, long-term anti-corrosion coating should be used, otherwise it easily leads to corrosion of other parts, and might affect the rigidity of the steel beam, and eventually lead to the degradation of the performance of the bridge reinforcement.

The connecting screw adopts a ribbed steel bar. The grade of steel bar is not less than HRBF400. The diameter of the steel bar is not less than 16 mm, and the spacing is not more than 0.5 m.

The U-shaped steel hoop meets the following requirements: (1) It is welded with a steel plate with a thickness not less than 6 mm; (2) It is made of Q345 or Q235 steel; (3) It can be made of weather-resistant steel, carbon steel or low-alloy high-strength steel; when the latter two are used, the exposed surface should be coated with long-term anti-corrosion coating; (4) When the steel plate at the diaphragm is bent, it should be wrapped around the outer surface of the diaphragm and fixed with anchor bolts; (5) The internal surface of the U-shaped steel hoop should be shot-blasted to an Sa2.5 level in the factory. If rust is found on site, shot blasting should be conducted manually to an St3.0 level.

The pull-up screw needs to meet the following requirements: Ribbed reinforcement should be used. The grade of reinforcement should not be less than HRBF400.

The diameter of the reinforcement should not be less than 16 mm, and the spacing is 0.5–1.0 m.

After the U-shaped steel plate hoop is hoisted in place by sections, it is welded as a whole. This method is an improvement for the heavy and difficult hoisting of the U-shaped steel hoop. The U-shaped steel hoop can be divided into several pieces suitable for hoisting, then hoisted in turn, and finally welded and fixed after being put in place.

The U-shaped steel hoop extends under the top plate of the T-beam or the small box girder, and the U-shaped steel hoop and the top plate are filled with high-strength bonding material. When the fire-damaged area is small and the damage to the top plate of the T-beam or the small box girder is not serious, only the loose and peeling concrete can be chiseled out, and then the peeling part can be repaired with polymer repair mortar. When the roof damage exceeds a certain degree, the U-shaped steel hoop which has covered the web needs to continue to extend, the lower part of the roof should be wrapped further, and then filled with high-strength bonding material to improve the integrity.

The method of strengthening can greatly improve the vertical stiffness and integrity of the bridge damaged by fire, and recover the operational bearing capacity of the bridge as soon as possible. At the same time, the method of reinforcement without brackets can not only solve the problem of insufficient vertical stiffness and horizontal integrity of this kind of bridge, but it can also solve the problem of not being able to erect brackets or the high cost of erecting brackets under the bridge. At the same time, it has the advantages of a short time-consuming reinforcement process and suitable construction.

**Problem 5.1.10**

In case of a large collapsed cavity of a mountain tunnel and low requirement of ground deformation, a pipe shed support and insert plate can be used to form a scaffold on the outer edge of the tunnel as support, while the collapsed cavity is filled with light materials to control the expansion and falling of the collapsed cavity, and reduce the supporting load to limit the adverse effect of the collapsed cavity. Figure 5.15a shows the collapsed situation of the Yongjia Tunnel (Fig. 5.16). In addition, when the tunnel was driven to the K16 + 183 tunnel face, a collapse of the tunnel top was caused. The longitudinal width of the collapse was about 11 m, the transverse width was about 10 m, the height was 22 m, and the volume was estimated to be about 2420 $m^3$. By means of a surface geological survey and observation, it was found that there was a concealed fault structure near the working face of mileage K16 + 183, which struck to the northeast and inclined to the northwest. There were many broken intercalations in the rock and some of them were thick. The stability of the rock in the broken section was poor. The remaining thickness of the collapse was 4.0 m, and the thickness of the collapse cavity was 3.5 m. The lower part wa the accumulation of the collapse, with the roof falling.

**Solution** According to the principle of energy, the engineering measures are essentially to ensure $\Delta U > \Delta T$. The way is to form a scaffold through the pipe shed

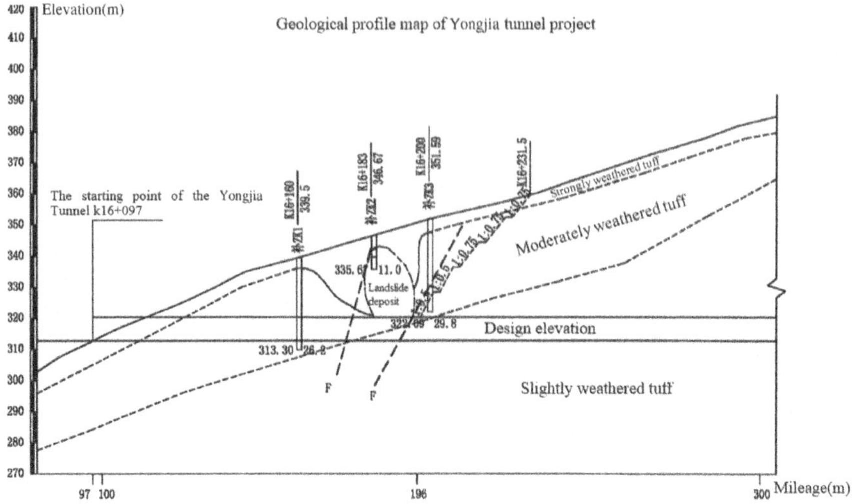


**Fig. 5.16** Filling light materials in the collapsed cavity of the Yongjia Tunnel to control the expansion and falling of the collapsed cavity

support and the insert plate, to play the role of support, and to improve the working $\Delta U$ of the resistance of the system. Filling light materials controls the expansion and falling of the collapsed cavity, reduces the support load, reduces the load work $\Delta T$ ($\Delta T + P\Delta S_1 + W\Delta S_2$), and prevents the adverse force and energy from transferring or concentrating to the weak part of the structure.

The tunnel collapse treatment plan adopts a pipe shed through the landslide area, filled with foam concrete in the cavity above the pipe shed, and the concrete operation will be carried out after the concrete strength reaches 70%. After being filled with foam concrete, the broken rock mass in the cavity can be prevented from collapsing, and the load above the pipe shed can also be reduced to ensure the safety of the surrounding rock and the supporting structure.

**Problem 5.1.11**

The original design of a tunnel is 390 m long with two lanes and 362 M long with three lanes, as shown in Fig. 5.17a. Influenced by multiple factors, such as structural belt mountain fragmentation, coal seam goaf, bias pressure, short distance, typhoons and heavy rainfalls, at 20:35 p.m. on August 22, 2013, the tunnel mountain and the left tunnel cave collapsed, and the right tunnel collapsed at 23:10 p.m., among which the mountain slipped down about 15 m along the sliding surface. The process of collapsing occurred as follows: The first collapse of the left tunnel → unstable collapse of the mountain → collapse of the right tunnel → continued instability of the mountain until it was relatively stable, as shown in Fig. 5.17b, c. According to the difference between the current terrain and the original terrain, the surface cracks and other factors, combined with the analysis of the process of collapsing, the collapsed

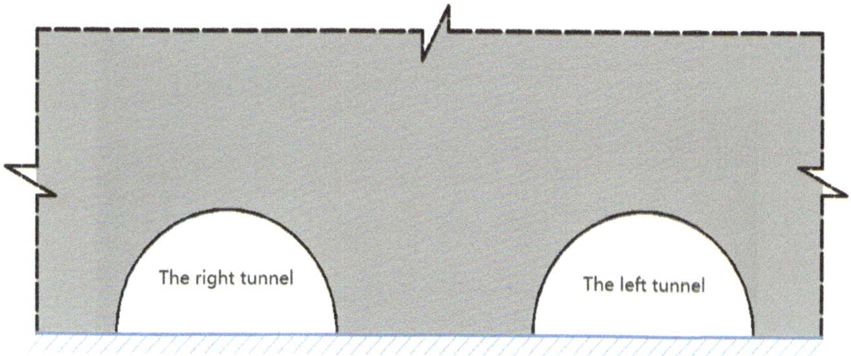


(a)

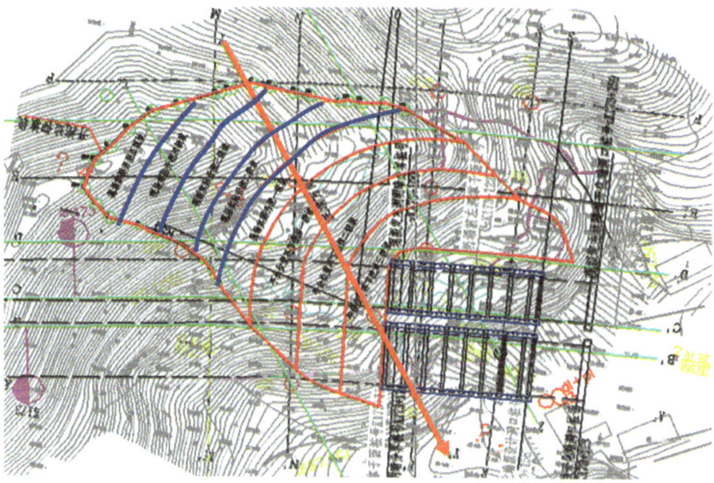

(b)

**Fig. 5.17** **a** Cross-section layout of a tunnel; **b** Layout plan of a tunnel portal and the direction of the collapse; **c** The process of the collapse of a tunnel portal

(c1)

(c2)

**Fig. 5.17** (continued)

body had the characteristics of first collapse and then the sliding, and the scale was medium.

Analysis of the process of collapsing: (1) The upper stage footage of the left tunnel was 9 m (ZK173 + 151), and the middle stage had just been excavated, and the front slope of the left tunnel had cracks, which indicated that the initial support was weak and not timely formed, and did not resist the pressure of the mountain slope of the portal. (2) The initial support of the left tunnel excavation reached ZK173 + 140.5, the arch foot cavity of the retaining arch was filled, the steel support was set in the initial support of the tunnel, and cracks appeared in the arch and the initial support of the tunnel, which indicated that the initial support of the left tunnel was weak, the footage was too long, there was no ring in time, and it could not resist the pressure and thrust of the mountain slope of the portal. (3) The first branch of the right tunnel YK173 + 146–YK173 + 137 was excavated. Remember that when the left tunnel is not completely stable and safe, it is not appropriate to excavate the right tunnel, which would have harmful effects on each other. (4) Continuing to excavate and support the right tunnel, the harmful deformation exceeded the limit, and so remember the lesson. (5) Typhoons and heavy rainfalls aggravated the pressure and thrust of the cave slope, causing the collapse of the left and right caves successively.

**Solution** Based on the original design and construction measures, the following measures were added to facilitate the stable equilibrium and the control of the deformation coordination of the portal slope and tunnel excavation support. (1) The loose mountain slope of the portal may be within the limit of the state of equilibrium. It was found that the cavity should be backfilled by grouting in time and the surrounding mountain slope of the portal should be provided with an anchor cable frame to facilitate the control of bearing the mountain pressure and thrust, so as to improve the stability of the mountain slope of the portal. (2) In the early stage, a double-layer initial support should be considered, and the inverted arch should be closed in time for a short excavation, so that the initial support bears most of the mountain pressure and local thrust.

According to the analysis of the process of the collapse of the tunnel portal, the treatment measures (Fig. 5.18) should be as follows: (1) The slope protection should be made from the top to the bottom of the existing collapsed body, and the elevation of the slope bottom should be the top of the crown beam of the open pit support pile or the top of the open pit back pressure backfill. The slope should be supported along with the excavation, and the excavation step distance should be controlled at the height required for the construction of a row of anchor bolts (cables). In addition, close attention should be paid to the geological conditions of the exposed surface during the excavation, and a small conduit grouting should be added if necessary. (2) A 39 m open cut via the foundation pit method and a 38 m large excavation open cut should be set for the left and right tunnels respectively. There were 60 foundation pit retaining piles with a length of 22.8 and 25 m. It was planned to use percussion drilling to make holes. (3) After the construction of the foundation pit retaining pile has been completed, the pouring of the top crossbeam, the crown beam and the diagonal bracing should be started. The earthwork of the right tunnel foundation pit should be excavated first, and the foundation pit enclosure should be provided with 4 rows of supports, the upper two rows being concrete supporting beams, the lower two rows being temporary steel pipe supports, and the excavation of the first level of support construction is the first level. (4) After the completion of the open cut construction via the foundation pit method of the right tunnel, the open cut backfilling of the right tunnel should be carried out in time. After the completion of the backfilling, the open cut construction of the foundation pit section of the left tunnel should be started, and the concealed tunnel construction of the right tunnel

**Fig. 5.18** Reinforcement and excavation support of a tunnel portal slope

should be started as well. The second lining of the right tunnel is to be poured to 50 m and the excavation of the left tunnel can be carried out until it is through.

## 5.2 Equilibrium and Stability Theory of Underground Engineering

According to statistics, all kinds of accidents involving a collapse are highly related to the safety of the engineering structures. The proportion of deaths in accidents is: temporary facilities and support collapse accounts for 32.6%, underground engineering construction collapse accounts for 32.6%, foundation pit excavation and retaining wall collapse accounts for 23.9%, road and bridge collapse accounts for 9.9%; among the roads and bridges, the proportion of dangerous bridges is high, and the potential safety hazard cannot be ignored: the service life of a bridge is within 10 years, the proportion of bridges in danger accounts for 20, 24% in 10–20 years, and 20% in 20–30 years; even in developed countries in Europe and America, the structural defects of roads and bridges are as high as 11%. The basic reason for the high incidence of the above engineering problems is that there is a question of metastable equilibrium of the traffic engineering structure: The rate of collapse during the process of the construction of temporary structures and underground engineering structures accounts for 89.1%, which is mainly due to the instability of the branch point (the state of stress deformation of the structure changes, and the analysis of the design calculation is not consistent with the actual situation), and the high rate of collapse is normal; the rate of collapse of road and bridge structures is 9.9% in use and

64% in 30 years of use, mainly due to the influence of the cyclic load and overload, and the rate of collapse is relatively low in construction. Taking the construction of bridge and mountain tunnels as an example, it can be illustrated as follows: (1) The calculation of the design stress and deformation of each step of the construction of the bridge is basically consistent with the actual conditions of construction stress and deformation, except for non-human errors, there are fewer accidents involving collapses in the construction of bridge structures; (2) In the design of mountain tunnels, the status of the stress and deformation of the mountain tunnel structure is calculated and analyzed by unit length (i.e. 1 m). There is a gap or even a big difference between the status of the stress and deformation of the mountain tunnel structure in the design construction scheme or the standard guidance construction scheme or even the actual construction scheme and the original design calculation and analysis. When the condition of the surrounding rock is good, the problem is not obvious, and when the the condition of the surrounding rock is poor, the problem is sudden. The rate of the collapse of underground engineering constructions is about 56.5%. In many cases, instability of the branch point occurs in the state of structural stress deformation during the construction of mountain tunnels (the state of structural stress deformation has changed, and the original design calculation and analysis are inconsistent with the actual situation). The above situation should cause people to wake up! The code should be modified to guide the construction scheme, so that the state of the stress and deformation of the mountain tunnel structure in the design construction scheme or even the actual construction scheme should be basically the same as the original design calculation and analysis, so as to avoid the problems of collapse such as the instability of the branch point in the construction of the mountain tunnel.

In addition, according to the statistical results of the height of the collapse in the construction of the Chengdu Kunming railway and other tunnels, the former Ministry of Railways is close to 2 times the span value specified for shallow tunnels. Therefore, it is of practical significance to calculate the load of tunnel lining structures by Protodyakonov's theory in the code of tunnel designs no matter whether the tunnel is shallow or deep, and it is of the same significance to calculate the load by Protodyakonov's theory for metro shield in the shallow soft soil city. It is also a record that only lining cracks or water leakage are found in a well-constructed tunnel or subway, and few accidents involving lining structure damage have been recorded.

I. Practice is the foundation and source of theoretical innovation

   The author has been engaged in the research, construction and management of bridge tunnels for more than 30 years, since 1983. In the long-term process of work, the author has found that tunnel construction safety accidents and tunnel design theories are a pair of contradictions that have been perplexing the engineering and academic circles. At the end of 1992, the author chose many construction sites of the Nanning-Kunming Railway, participated in the construction of more than 100 tunnels along the Nanning-Kunming Railway, got in touch with a large number of front-line tunnel design, construction and management personnel, collected a large number of data on tunnel geology and changes in the state of

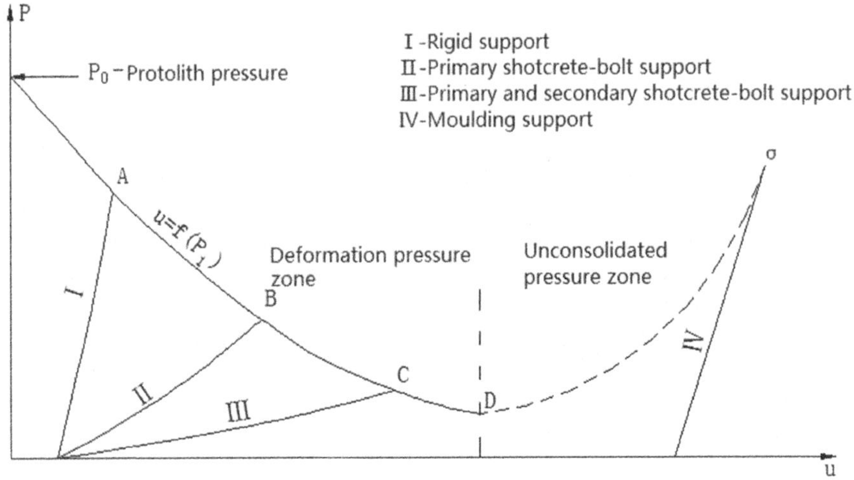


**Fig. 5.19** Displacement supporting characteristic curve of the surrounding rock (a general theory proposed in the process of solving specific underground engineering construction problems, generally corresponding to equilibrium and destruction, and may imply risks)

construction, and accumulated valuable experience regarding design, construction and management. These experiences have accumulated a lot for the author to think about and have given him a basis for putting forward the theory of equilibrium and stability.

In the subsequent work practice, the author realized that the applicability of NATM (Fig. 5.19) is related to the understanding and mastery of tunnel design theory by design and construction personnel. Through a large number of comparative analyses of the geological environment, construction management, construction equipment and materials, and form of the organization of the personnel of tunnels at home and abroad, it was found that "the traditional underground engineering theory generally corresponds to equilibrium and destruction, and the state of equilibrium may be a safe state or a state with hidden risks."

More than 100 bits of tunnel construction data and a large number of on-site technical discussions on the construction of the Nanning-Kunming Railway have accumulated rich material for the establishment of an applicable and safe theory of tunnel construction. In 1995, when the author began to analyze and sort out these data, he came up with some preliminary ideas. At the same time, he began to think about refining and sublimating

On the basis of inheriting the traditional theory according to the actual needs and combining the suggestions of experts in the fields of tunnels, mechanics, geology, materials, machinery, etc., so as to realize the innovation of a theory of tunnel design and construction, to provide new ideas for solving the new problems that have emerged in practice better.

In 1996, during the construction of the Wangjiashan tunnel of the Shimen-Changsha Railway, according to the condition of the tunnel passing through the

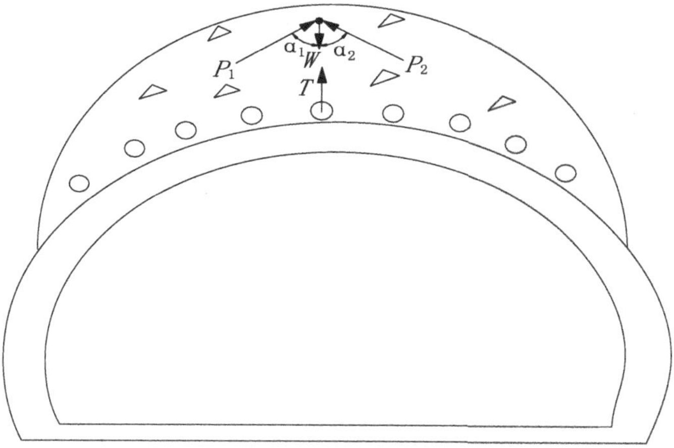


**Fig. 5.20 Law of the load transfer of the surrounding rock controlled by the layer of the effective bearing structure of a tunnel** $P_1\cos\alpha_1+P_2\cos\alpha_2+T = W$ Where, $P_1$, $P_2$—mutual supporting force of the surrounding rock; W—gravity; T—support resistance (T plays the role of combining $P_1$ and $P_2$)

ancient river bed, the author put forward the law of controlling the load transfer of the surrounding rock by the effective structural layer of the tunnel (see Fig. 5.21), that is, when the surrounding rock of the tunnel adopts a strong pre-supporting element and keeps the surrounding rock in its original state, it can give full play to the self-bearing capacity of the surrounding rock.

$$P_1\cos\alpha_1 + P_2\cos\alpha_2 + T \geq W \quad (5.2.1)$$

where, $P_1$, $P_2$—self-bearing force of the surrounding rock; W—gravity; T—support resistance (the support resistance T is as small as possible).

It can be found through an in-depth understanding of the formula (5.2.1) that: under the guidance of the concept of "reasonable exertion of the self-supporting capacity of the surrounding rock" and "basic maintenance of the original state of the surrounding rock", various reasonable construction methods, appropriate control of the support and process (based on the first two and the last three) should be adopted to make the interaction between the surrounding rock and the support reach stable equilibrium and control of the deformation coordination to ensure stress safety. In fact, only when the condition of "basically maintaining the original state of the surrounding rock" is met, can the self-supporting force $P_1$ and $P_2$ of the surrounding rock be as large as possible and the support resistance t as small as possible, that is to say, the self-supporting capacity of the surrounding rock can be reasonably developed.

**In a word, the physical meaning of formula (5.2.1) and Fig. 5.20 is based on the coordinated control of structural deformation, which contains both mechanical equilibrium and mechanical transfer and adjustment problems.**

**The force value of formula** (5.2.1) **and Fig. 5.20 on the supporting structure is less than that of pure mechanical equilibrium on the supporting structure, which is also the progressive significance of formula** (5.2.1) **and Fig. 5.20 on the underground engineering construction.**

During the construction of the Xiaodejiang tunnel of the Nanning-Kunming Railway in 1993 and Chenjiashan tunnel of the Shimen-Changsha Railway in 1997, we found that the small displacement of the mountain made the lining of the tunnel cracked. This makes us realize that the traditional theory of underground engineering (the theory of the bearing capacity of the surrounding rock and the theory of the loose load) and the engineering method (including NATM, the mining method, etc.) only solve the problem of an interaction equilibrium between the rock surrounding the tunnel and the support, but not the stability of the rock surrounding the tunnel and the support and their environment. Therefore, it is necessary to comprehensively study the stability of the mountain in the tunnel area and the stability of the interaction between the surrounding rock and the supporting system of the tunnel, so as to truly solve the control of the stable equilibrium and the deformation coordination of the tunnel structure and realize the safety of the stress of the tunnel structure.

In 1993, the engineering geological environment involved in the construction of the Chajiang tunnel of the Nanning-Kunming Railway was very complex. The surrounding rock was shale containing kaolin, which was easily expandable and led to the cracking of the tunnel lining. This makes us realize that tunnel construction needs not only objective management but also control of the process, and the supporting structure should be strong enough.

The experience of the construction of the Xiaodejiang tunnel and the Chajiang tunnel of the Nanning-Kunming Railway, the Wangjiashan tunnel and the Chenjiashan tunnel of the Shimen-Changsha Railway reminds us that the level of the underground engineering theory (the theory of the bearing capacity of the surrounding rock and the theory of the loose load) and the method of construction (including NATM, the mining method, etc.) must be improved to meet the needs of the design, construction and management of underground engineering in China.

From 1998 to 2008, the author studied the construction, altered design, accident treatment and management of more than 100 tunnels in depth, such as the Meiling tunnel, the Huangtuling (right lane) tunnel and the Baihe tunnel. During this period, the author's research team deepened the theory and received the careful guidance of the academicians Wang Mengshu, Liu Baochen and Sun Jun. On the basis of the existing achievements, we have put forward some comprehensive theories of the principle of the rationality discrimination of the tunnel construction scheme, the independent stress of the tunnel, the tunnel pre- support principle and the reasonable transfer path of force. In 2008, the author's research team further expanded the scale of its research, introduced the theory of structural equilibrium and stability and participated in the construction of shield tunnels such as the Qianjiang tunnel and the Hangzhou Metro Chenghu District, and finally preliminarily formed the theoretical framework of underground engineering equilibrium

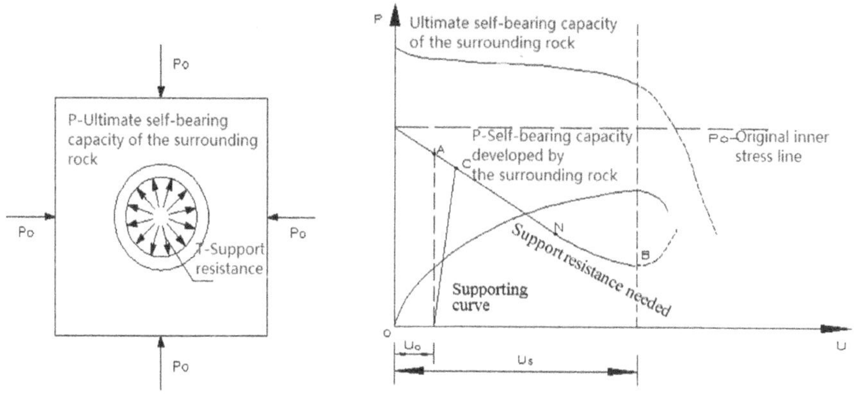


**Fig. 5.21** Force displacement characteristic curve of the theory of underground engineering equilibrium and stability (a theory proposed by the general mechanical model research problem, corresponding to the stable equilibrium and deformation coordination, which aims at eliminating hidden risks and puts it in a safe state)

and stability. In the past five years, we have increased the research on the classification of states of metastable equilibrium, the interaction of three kinds of states of equilibrium and Newton's equilibrium equation, until 2013, the theoretical system of underground engineering equilibrium and stability was basically formed (see Fig. 5.21). It should be said that this theoretical system derives from practice, which completes the accumulation of quantity, the evolution of quality, and the further theoretical sublimation and innovative development. It is the summary of more than 30 years' practical experience of the author's research team.

Compared with the traditional theory of underground engineering corresponding to equilibrium and destruction, the state of equilibrium may belong to the safe state or the state of hidden risks, the theory of equilibrium and stability corresponding to the control of the stable equilibrium and deformation coordination, eliminating hidden risks, and always belongs to the safe state.

According to the theory of equilibrium and stability, the basic requirements for maintaining the equilibrium and stability of underground engineering are as follows:

$$F = T + P \tag{5.2.2}$$

$$F > P_0 \tag{5.2.3}$$

The formula (5.2.3) is generally applicable to solving the problem of the equilibrium and stability of underground engineering. Corresponding to the coordinated control of stable equilibrium and deformation, if underground engineering meets the requirements of this formula, it can eliminate hidden risks and ensure safety of stress. The form of the theory of the equilibrium stability of underground engineering changes with the concrete form of the problem. Formula (5.2.3) shows that two basic concepts should be grasped in underground engineering, i.e. "reasonable exertion of the self-bearing capacity of the surrounding rock" and "basic maintenance of the original state of the surrounding rock". The theory of the loose load based on the traditional Terzaghi theory or Protodyakonov's theory is a mechanical method based on the statistical value of the shallow-buried loose stratum and the deep-buried loose rock and soil mass, and the state of the stress and deformation of the simple case naturally meets the "coordination and control of deformation". Therefore, there is no concept of deformation coordination and process control, nor the basic concept of "reasonably exerting the self-bearing capacity of the surrounding rock" and "basically maintaining the original state of the surrounding rock". For relatively broken rock or deeply buried strata, only when the condition of "basically maintaining the original state of the surrounding rock" is met, can the "reasonable self-bearing capacity of the surrounding rock" be achieved; the New Austrian Tunneling Method (NATM), the New Italian Tunneling Method (NITM), the Norwegian Method of Tunneling (NTM), the constraint convergence method and other rock-bearing theories are based on the basic integrity of the rock mass. Because the rock blocks in Austria, Switzerland and other European regions are relatively complete and have good mechanical properties, which can reflect the concept of "reasonably exerting the self-bearing capacity of the surrounding rock" and "basically maintaining the original state of the surrounding rock", the method based on the modern theory of the bearing capacity of the surrounding rock has good applicability in Europe. The eastern part of China lies at the junction of the Eurasian continental plate and the Pacific plate, while the western part of China lies in the Himalaya orogenic belt at the junction of the Eurasian plate and the Indian plate, with a complex and changeable geological environment. Under the influence of various complex geological processes, there is a complete rock mass, a relatively broken rock mass and a shallow buried loose stratum in China. If we copy the European theory and method completely, there will be deviation. Based on the concept of "reasonably exerting the self-bearing capacity of the surrounding rock" and "basically maintaining the original state of the surrounding rock", that is, "coordinated control of deformation", it is necessary to establish a generalized mechanical stable equilibrium equation suitable for various structural characteristics of rock and soil mass, and pay attention to the research of a reasonable excavation method, supporting structure measures and control of the construction process, to ensure that the interaction between the surrounding rock and the support can achieve a stable equilibrium and control of the deformation coordination. This is also the basic requirement of the analysis of the structural design and safety of underground engineering. For the shallow-buried tunnel in loose ground or the shield tunnel in soft soil, the simplified method of load structure (corresponding to the theory of the loose load) can be adopted, and the error is within the allowable range of strength of the supporting structure; however, the construction method and process control measures should adopt the thinking of the stratum structure method (corresponding to the theory of the bearing capacity of the surrounding rock), realize the joint action of stratum and support to achieve a "stable equilibrium and control of the deformation coordination", eliminate hidden risks and ensure safety of stress. No matter whether dealing with surface engineering or underground engineering, as long as the interaction with the rock and soil mass reaches the "stable equilibrium and control of the deformation coordination", the basic conditions for studying the "better solution" of the engineering structure can be expanded to "reasonably develop the self-supporting capacity of the rock and soil mass" and "basically maintain the original state of the rock and soil mass".

Therefore, although the structural mechanics method of the "theory of the loose load" (mining method, etc.) is clear, it is difficult to control the supporting structure in place, and many rock and soil masses cannot effectively form a stable combination system of the stress and deformation of rock and soil masses and the supporting structure; "the theory of the bearing capacity of the surrounding rock" (NATM, etc.) is based on hard rock or soft rock reinforced by advance support. Although the supporting structure is advanced and the concept of giving full play to the self-supporting capacity of the surrounding rock has been proposed, the structural mechanics method is not clear, and some rock and soil masses cannot effectively form a stable combination system of rock and soil mass and supporting structure under the same stress and deformation; generally speaking, in the past underground engineering construction, the layer of the bearing structure formed in advance or promptly and effective control and the measures for the control of the structural mechanical deformation which lack the guarantee of the space-time effect in the construction process are often ignored, and so it is difficult to control the unreasonable or even harmful transfer of the force, and the structure is prone to collapse.

Facing the construction problems such as collapse, water leakage and stratum deformation caused by underground engineering crossing a complex engineering environment and unfavorable geological conditions. On the basis of the traditional theory of the equilibrium and stability of engineering structures, according to the coordinated control method of deformation of engineering structures, combining the advantages of "the theory of the loose load" and the "theory of the bearing capacity of the surrounding rock", the author establishes the theoretical system and supporting key technologies of the equilibrium and stability of underground engineering. Based on the overall equilibrium and stability of the surrounding rock and supporting structure, the New Austrian Tunneling Method (NATM), the Norwegian Method of Tunneling (NTM) and the New Italian Tunneling Method (NITM) are classified and compatible. According to the different stages of bearing capacity of the surrounding rock, the existing achievements can be used to prevent and control the quality and safety of the project and ensure the quality and safety of the construction of underground engineering (see Fig. 5.22), which can be divided into three special cases.

Fig. 5.22 **a** NTM curve; **b** NATM curve; **c** NITM curve

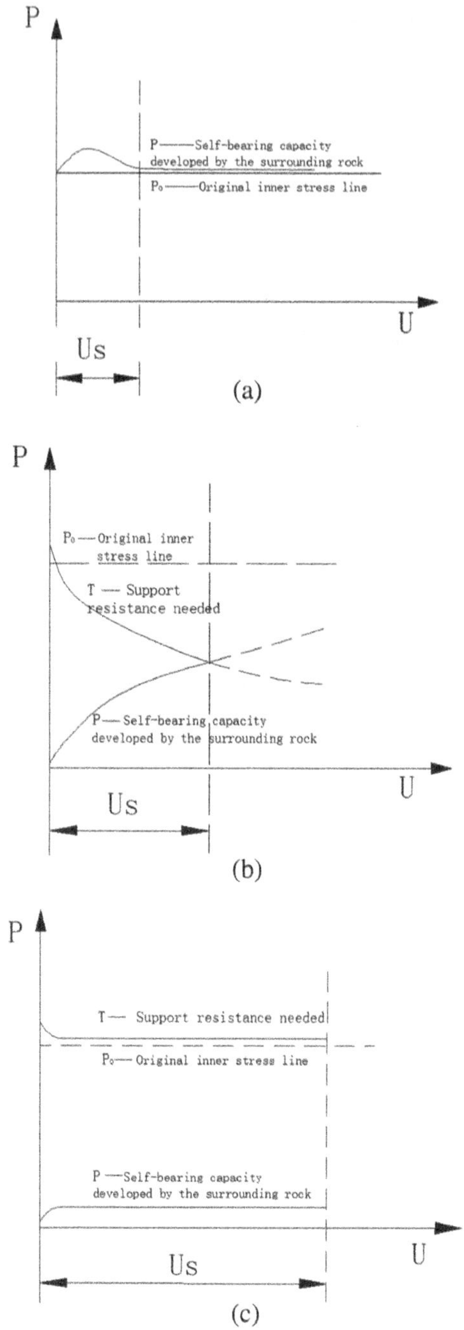

The "excavation energy control technology" [$E = E_1 + E_2 + E_3$], control blasting and excavation process to the surrounding rock and the energy consumption $E_2$ of the pre-supporting structure disturbance is the smallest, and using its choice of construction method not only provides a clear idea for the selection of an appropriate construction method in complex conditions, but it also realizes the minimization of the disturbance of blasting and mechanical excavation to the surrounding rock, ensures the stability of the surrounding rock and saves the cost of excavation; "Strong pre-support technology" [$F > P_O$ or $\Delta U > \Delta T$], the pre-supporting structure can basically maintain the original state of the surrounding rock, give full play to the self-supporting capacity of the surrounding rock, and avoid accidents due to excessive loose deformation or even collapse; "Comprehensive technology of independent stress" [$P_V = 2a\,(3a_1^2 - a^2)/(3a_1 f)$—approximate solution to the stress of the rock surrounding the tunnel is directly proportional to the square of the span], the technology for the control of the deformation coordination and technology for the control of the excavation energy are used to strengthen and guarantee the reasonable design state of stress and deformation in the middle column area of the tunnel, effectively solving the problem of interaction between the double arch tunnel and the small clear distance tunnel in the process of excavation and support; "Technology for the control of deformation " [$P + T \geq P_O$], through structural control measures to ensure the basic maintenance of the original state of the rock and soil mass (surrounding rock), effective control of any part of the rock and soil mass (surrounding rock) plays a role in balancing the system of the underground structure, transforming the load or load borne by the structure into resistance or resources that play a role in structural balancing, and reasonably playing the self-bearing capacity of the surrounding rock. Reduce the interaction between the rock and soil mass and the environment in the underground engineering construction, and ensure safety during the construction of the tunnel under the complex environment, such as loose accumulation, expressway and existing buildings in the city; compared with the three principles of "smooth blasting, shotcreting and anchoring support, monitoring and measurement" in the tunnel design and construction specifications, the four key technologies proposed by the project team are suitable in practical application and conducive to ensuring the construction quality and the safety of the underground engineering structures. Therefore, the system of the theory and the key technology of underground engineering equilibrium and stability reveal the essence of the stable equilibrium of the state of the stress and deformation of underground engineering, ensure the rationality of the structural design and construction technology of underground engineering and the correctness of the results of the mechanical analysis, avoid the complications, such as the transfer of the stress path and deformation incongruity of some underground engineering structures, and provide a decision-making basis for choosing a better design and construction scheme of underground engineering; it is also conducive to solving many complex underground engineering problems; practice has proved that most of the problems of construction failure of underground engineering are caused by the failure to grasp the essence of the stable equilibrium of the state of the stress and deformation of underground engineering, and the failure to prevent and control the engineering quality and safety according to the different stages of the bearing capacity of the surrounding rock.

II. The unity of the four levels of the theoretical system of the analysis of engineering structures

The theoretical system of the analysis of engineering structures includes four levels: (1) Basic mechanical laws (equilibrium and stability, reasonable exertion of the self-bearing capacity of the surrounding rock, etc.); (2) Mechanical concepts that are easy to understand and master (basic maintenance of the original state of the surrounding rock, coordinated control of deformation, reasonable transfer path of force, target control and process control, etc.). By using these mechanical concepts, the mechanical properties of engineering structures can be judged intuitively, which can verify the factors that can be easily ignored in the complex analysis or the deficiencies caused by the inertia of thinking, reduce the design or construction errors, and prevent the occurrence of disasters. By mastering these mechanical concepts, we can easily grasp the mechanical characteristics of the actual engineering structure in the process of design, construction and management, so that it is always in a state of stable equilibrium and control of deformation coordination, and ensure the safety and effectiveness of the engineering structure; (3) The basic method the process of design and construction are the means and guarantee to realizing the above laws and concepts; (4) The actual engineering condition and the scheme of the structural system design and construction (foundation condition, load condition, environmental impact, reasonable structure and structural system, method of construction, etc.) are the specific measures at the operational level.

The above four levels are unified in practice. It shows that the theoretical system of the analysis of the engineering structure is unified in the process of gradual improvement, which avoids the phenomenon of "from a high position but not from a low position", and embodies the general applicability of the strategy.

III. The universality and tactical applicability of the theory of engineering design and construction

From a macro point of view, various underground engineering theories, including the mining method, the NATM method, the shallow excavation method, the Norwegian Method of Tunneling (NTM), and the New Italian Tunneling Method (NITM), have little difference in mechanical nature, and these methods have a strategic guiding role and universality. The realization of these strategic intentions needs to be guaranteed by the reasonable methods of construction in the above-mentioned level (3), and at the same time, the specific measures in level (4) are needed in order to carry out the design and construction according to the actual engineering conditions. In the form of expression, various theories differ greatly, so we should analyze the specific problems. Therefore, various methods have tactical applicability.

Taking tunnel engineering as an example, throughout nearly two centuries of exploration, various design theories and methods of construction have been formed, such as NATM, the shallow excavation method, the mining method, the theory of the equilibrium and stability of underground engineering, etc. and these design theories and methods of construction have played a very important role in the practice of tunnel construction. In fact, there is no theory or method of construction that can be applied to any tunnel construction. When some design theory and method of construction are not applicable to the specific practice of some tunnels, it shows that the theory and the method of construction have defects and need to be improved.

A good theory cannot only explain the mechanical behavior of the tunnel, but it can also promote the progress of engineering construction. However, each theory and method of construction is limited by its time and historical conditions. On the premise of following the basic mechanical laws, specific problems should be analyzed to achieve the unity of strategic universality and tactical applicability of the theory of the engineering design and construction.

IV. Popularization of theory facilitates the combination of the control of the process and of the target

It is the category of objective management to adhere to the basic mechanical law or essence of level (1) (such as equilibrium and stability, giving full play to the self-supporting capacity of the surrounding rock, etc.) so that the practice of engineering design and construction will not deviate from the correct direction. By analyzing the forms of expression of various laws, the mechanical concepts that are easy to understand and master (such as basic maintenance of the original state of the surrounding rock, the control of the deformation coordination, reasonable path of transfer of the force, the control of the target and of the process, etc.) are extracted. Using these mechanical concepts to judge the mechanical properties of engineering structures directly can reduce engineering errors and prevent catastrophes, which is the category of process control and management. These popular concepts are suitable for designers and constructors to be able to grasp in the design, construction and management of the actual system of the engineering structure. Therefore, it is beneficial to popularize and master the profound theory expressed by popular basic concepts. Through the popularization of theory and the combination of the control of the engineering process and the control of the objective, the problems can be solved in their bud, and the remedial measures can be taken after the application process of construction and operations in order to avoid the problems, so that the actual system of the engineering structure is always in a state of stable equilibrium and control of deformation coordination. For example: Fig. 5.23 shows that before the reconstruction of the Pailong section of Sichuan-Tibet Highway, there were many dangers, which could easily cause traffic accidents; Fig. 5.24 shows that through the reconstruction project of four tunnels and two bridges of the system, the hazards such as mountain collapse, river water damage and debris flow were prevented and the safety of the traffic along the highway has been guaranteed.

**Fig. 5.23** Before the reconstruction of the Pailong section of the Sichuan-Tibet Highway, there were many dangers, which could easily cause traffic accidents

**Problem 5.2.1**

The new type of quantitative yielding and energy dissipation connecting device and its application to the tunnel expansion surrounding rock arch often face a great degree of deformation of the surrounding rock, rock bursts, collapses, water inrush and mud inrush and other disasters during the process of the construction of the tunnel and

**Fig. 5.24** The Parlung Bridge, tunnel and other projects prevent and control the hazards, such as mountain collapse, river water damage and debris flow to ensure the safety of the traffic along the road

underground engineering, especially in the deep underground space and complex stratum.

In order to prevent or avoid these disasters, experts and scholars at home and abroad have done a lot of research and gained practical experience on the support of underground structures. The current arch forms mainly include: 1. The U-shaped steel arch support; 2. The steel tube confined concrete arch support; 3. The yielding bolt support; 4. The yielding arch support, etc. The U-shaped steel arch support is the most commonly used passive support method in the soft rock roadway. It directly acts on the surrounding rock surface by providing passive radial support force to put the deformation pressure of the surrounding rock in equilibrium and restrict the deformation of the surrounding rock. However, there are many problems in the support of the U-shaped steel arch: ① The supporting strength is insufficient; ② The working resistance is low; ③ The support capacity is not fully developed, which seriously reduces the utilization rate of materials, and the effect of support is not guaranteed; ④ The quantitative yielding is not allowed; the steel tube confined concrete arch is a supporting form with a high bearing capacity, certain compressibility, convenient construction and good mechanical properties. However, there are also the following problems: ① The contact area between the circular section of the concrete-filled steel tube arch and the surrounding rock surface is small, which can easily create a concentration of stress; ② The cost is high; ③ There is a lack of special underground installation equipment; ④ The weight of the arch is great; ⑤ There is no quantitative yielding function; the uncertainty or error of the yielding bolt support from the stable rock layer will make the operators feel unsafe; in the past decades of engineering practice, people have made a lot of supporting materials and supporting structures with yielding function, and have improved the original rigid supporting structure to increase the contraction of support. However, most of the yielding arch supports used in underground engineering only consider the deformation, and the yielding arch supports can only carry out the free type of yielding, and it has not been reported that the quantitative yielding and energy dissipation connecting device of the steel structure can be realized in the work of arch support.

**Solution** Provide a new type of quantitative yielding and energy dissipation connecting device and method of manufacturing which is suitable for the underground engineering arch support featuring a simple structure, convenient installation, and controllable yielding time and amount, especially for the realization of a yielding connecting device of the arch support of various deep, soft rock and other underground engineering elements, such as roadways, tunnels and cavern groups, which are difficult to support due to a large deformation. The technical scheme adopted to solve the technical problem is: A new type of connecting device for quantitative yielding; this device includes a yielding core member, a yielding core sleeve, an upper connecting device, a lower connecting device, an upper top plate, a lower bottom plate, a joint connecting device, a sliding hole I, and a sliding hole II, an upper connecting bolt, a lower connecting bolt, an upper arch structure plate supporting the main structure, and a lower arch structure supporting the main structure; wherein the yielding core component is fixedly connected to the upper top plate, the yielding core sleeve is fixedly connected with the lower bottom plate, the upper connecting device is connected to the upper top plate, the lower connecting device is connected to the lower bottom plate, the upper connecting device is provided with a sliding hole I, and the lower connecting device is provided with a sliding hole II. The joint connecting device is fixed in the sliding hole I of the upper connecting device and the sliding hole II of the lower connecting device. The upper connecting device and the lower connecting device are closely connected by the joint connecting device. The upper top plate which is connected to the upper connecting device and the yielding core component and the lower bottom plate which is connected to the lower connecting device and the yielding core sleeve are assembled to form a box structure of the new quantitative yielding and energy dissipation connecting device. Then, this box structure is connected to the upper arch structural plate and the lower arch structure of the main supporting structure by means of the upper connecting bolt and the lower connecting bolt, as shown in Fig. 5.25.

At present, the problems of the failure of the early support and even the safety of the existing squeezed large deformation surrounding rock tunnel are pretty prominent. It is difficult to solve the above problems by means of the traditional design and method of construction of the excavation support. It has been proved that: (1) The deformation of the surrounding rock can be reduced by a short excavation and the close support in time under the condition that the tunnel face is basically stable; (2) After excavation, the initial deformation of the surrounding rock tunnel with a great amount of deformation is inevitable, and the stability and yielding support can control a certain amount of deformation of the surrounding rock, which is conducive to the reasonable development of the bearing capacity of the surrounding rock; (3) Under the condition of there not being enough radial support strength, it is difficult to form the self-supporting structure of the surrounding rock with a great deal of deformation, and the supporting structure can also easily fail in its function, so it is difficult to form a stable and effective bearing structure for the surrounding rock; the arch frame with a certain degree of rigidity and strength is an effective way to solve the problem. Therefore, the reasonable method of excavation, primary yielding support and secondary rigid support are the effective ways to give full play to the self-bearing capacity of the surrounding rock with a great deal of deformation (Fig. 5.26).

**Problem 5.2.2**

With the expansion of coastal cities, highway construction needs to pass through the tidal bore area, and the inner side of the highway needs to fill the sea and create land for the development of the new area. The bridge crossing the tidal bore area will affect the new area and the marine landscape. The new area plans to use highway tunnels to cross the tidal bore area. Shallow buried highway tunnels crossing the tidal bore area is not only conducive to land reclamation, but also conducive to the longitudinal slope of the highway. How can this be designed and constructed in order to ensure the safety of the tunnel construction and eliminate the influence of the tidal bore?

**Solution** In view of the existing new tunnel, there is a need to be buried in a shallow manner through the tidal bore area of tideland construction technology; for example, the construction technology of shallow-buried subsurface excavation and pipe sinking cannot meet the requirement of a shallow-buried highway tunnel crossing the tidal bore area, which is not only conducive to land reclamation but it is also conducive to the longitudinal slope of the road; however, it is difficult to prevent and control the leakage of the tidal bore in the tunnel area by using shotcrete

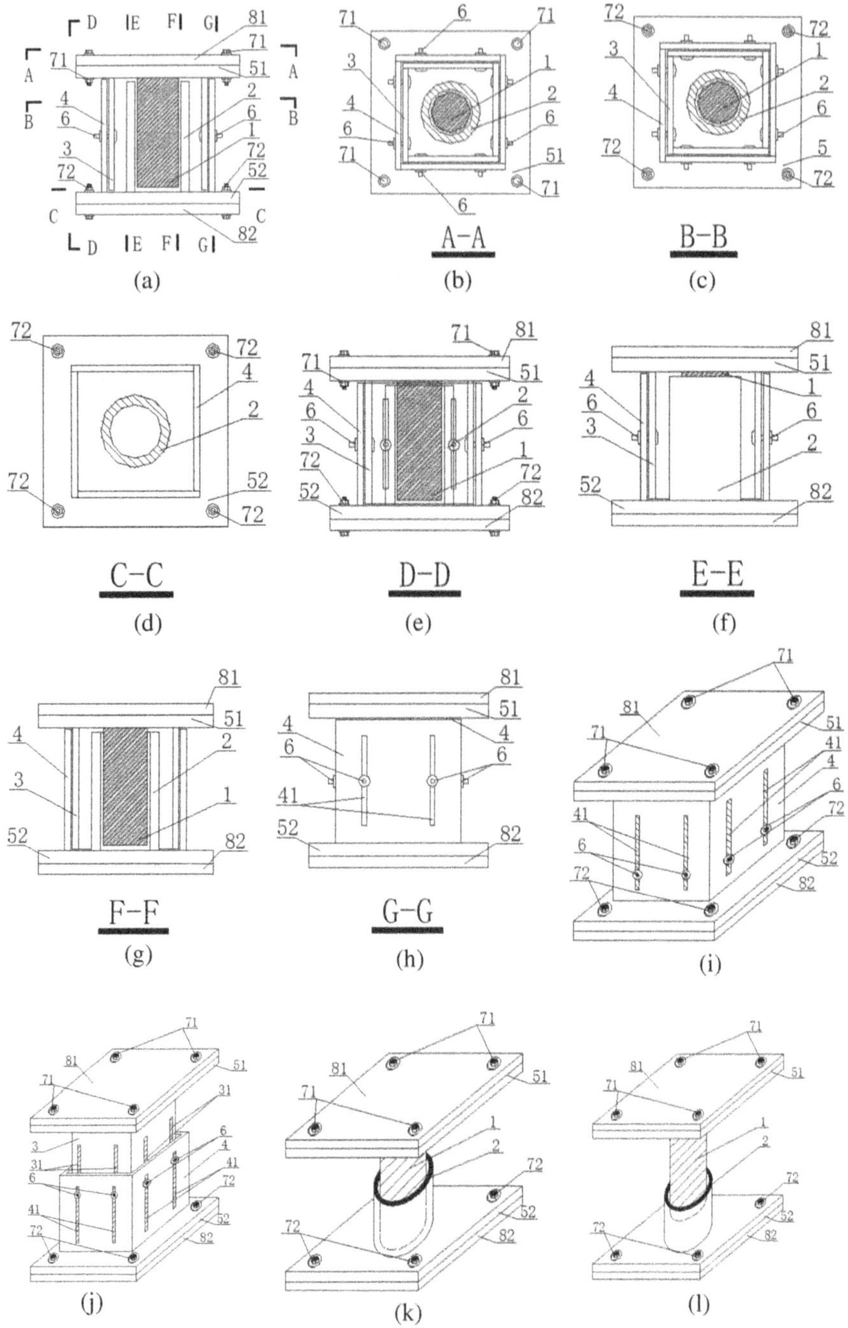


**Fig. 5.25** A new type of connecting device for quantitative yielding and energy dissipation of arch support for the expansive surrounding rock of a tunnel

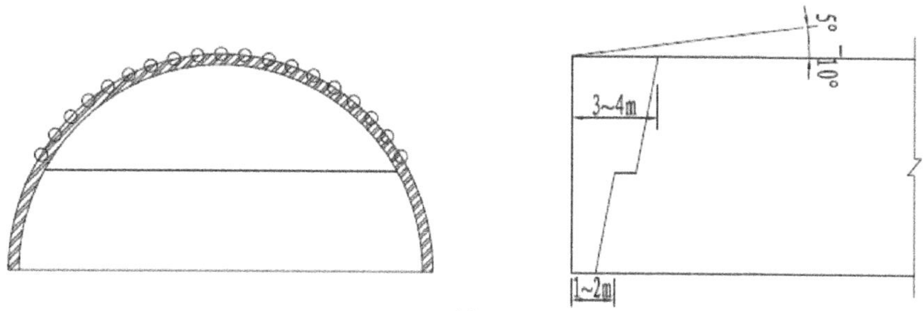


(a)

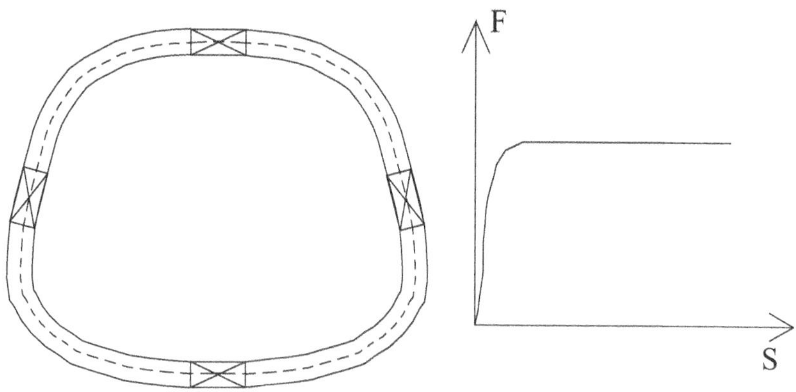


(b)

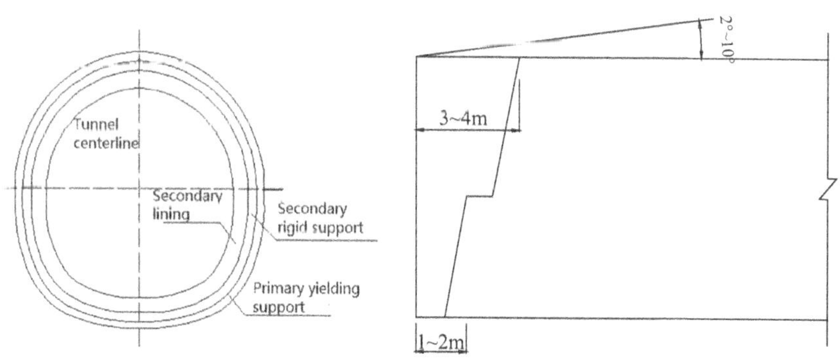


(c)

**Fig. 5.26** **a** Diagram of the approximate full cross-section method of construction of the short step reserved core soil; **b** Schematic diagram of the yielding arch and its law of stress and deformation; **c** Diagram of the support and construction of a tunnel with squeezed surrounding rock with a great deal of deformation (unit: m) at the side of the foundation pit and a common cushion at the bottom of the tunnel, which has a direct impact on the cast-in situ quality and structural durability of the reinforced concrete in the lower half of the tunnel. Combined with the experience of reclamation and land making in the water conservancy department, the government first entrusts the water conservancy department to building a tidal bore dam in accordance with the design line of the highway tunnel, and then sets up a reliable structure in the tunnel area to prevent tidal bore leakage and uneven settlement and cancels the optimization measures such as uplifting piles. Finally, the tunnel structure is constructed according to the design requirements, so as to ensure the quality and safety of the tunnel construction and eliminate the impact of tidal bore (Fig. 5.27).

**Problem 5.2.3**
It is about the method of construction of the approximate full cross-section for the approximate critical stability (including broken) of the surrounding rock of the highway tunnel in mountainous areas. The approximate critical stability (including broken) of the surrounding rock of many highway tunnels in mountainous areas in China are constructed by means of the CD/CRD method, the multistep method or multiple combination methods, as shown in Fig. 5.28a. However, the different length of longitudinal excavation and support in different parts will lead to a lack of space stability in the process of tunnel construction or the difficulty of forming an effective layer of the bearing structure. On the other hand, the design of the tunnel is a plane mechanical problem, while the construction of partial excavation and support is a space mechanical problem. There are differences between the design of tunnels and the construction of partial excavations and supports. When the condition of an approximate critical stability (including broken) of the surrounding rock of the tunnel is poor, tunnel construction collapse and even casualties could often occur.

**Solution** In view of the contradictions between the existing design of the approximate critical stability (including broken) of the surrounding rock of mountain highway tunnels and the mechanical problems in the construction of partial excavations and supports, when the condition of the approximate critical stability (including broken) of the surrounding rock of the tunnel is poor, tunnel construction collapse and even casualties could often occur. By shortening the longitudinal length of the construction of the tunnel excavations and supports, the construction of the tunnel excavations and supports is close to the plane mechanical problem. In the practical construction of the approximate critical stability (including broken) of the surrounding rock of the highway tunnel in mountainous areas, two effective approximate full cross-section construction methods for the approximate critical stability (including broken) of the surrounding rock are formed.

(1) When the tunnel face is basically stable during the process of construction, it is advisable to adopt the construction method of the lower pilot tunnel in advance of the full cross-section, as shown in Fig. 5.28b. For class III, IV and V surrounding rock tunnels with strong pre-supports, the lower pilot tunnel is adopted to advance 3–5 m, and the rest of the approximate full cross-section

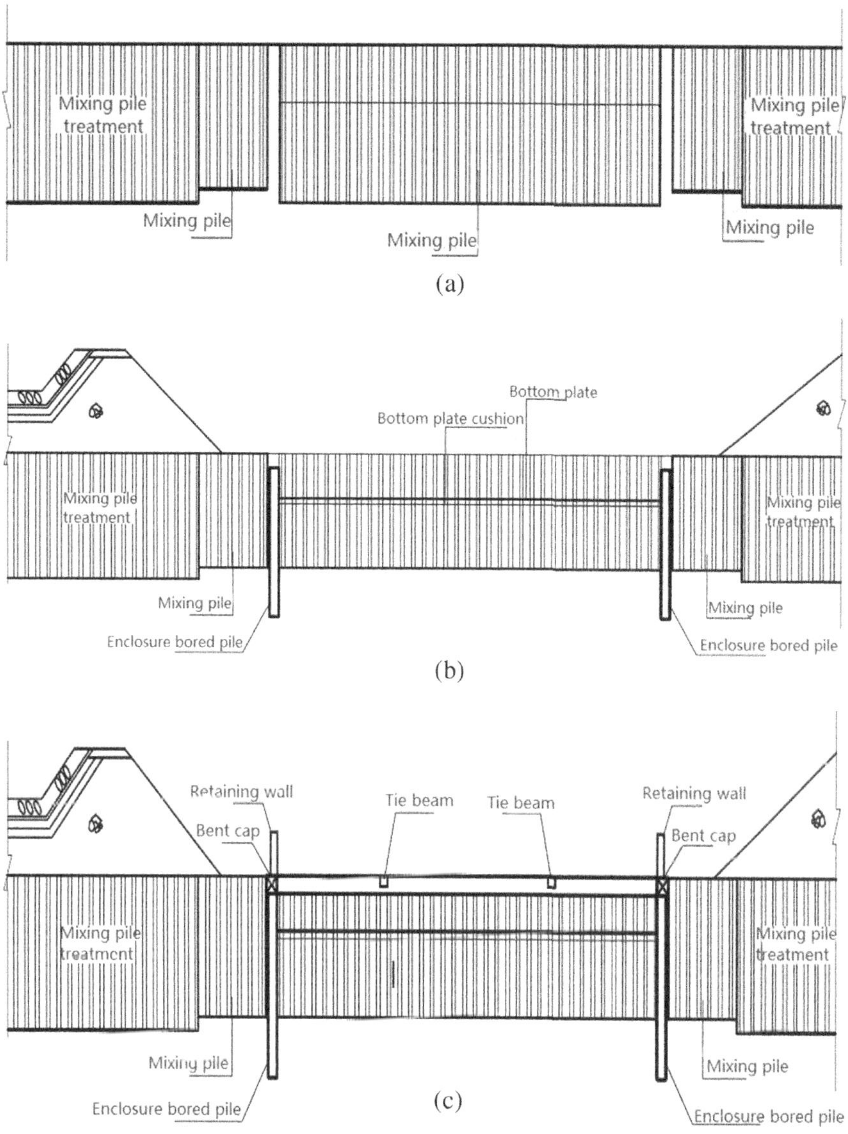


**Fig. 5.27** **a** Construction drawing of tidal bore dams on both sides and treatment of the subgrade for preventing tidal bore leakage and uneven settlement in the tunnel area; **b** Construction drawing of tidal bore dams on both sides and enclosure bored piles preventing lateral displacement and leakage in the tunnel area; **c** Construction drawing of inserting a prefabricated structural slab and crown beam and other supports and retaining walls which guarantee construction safety behind the enclosure bored piles on both sides of the tunnel area; **d** Construction drawing of the construction access road on both sides of the tunnel area; **e** Construction drawing of the prefabricated reinforced concrete structure of a foundation pit and bottom plate in the tunnel area; **f** Construction drawing of the tunnel floor structure; **g** Construction drawing of the temporary support removal and support replacement; **h** Construction drawing of the upper main structure of a tunnel; **i** Construction drawing of the symmetrical backfilling and landscaping on both sides of the tunnel

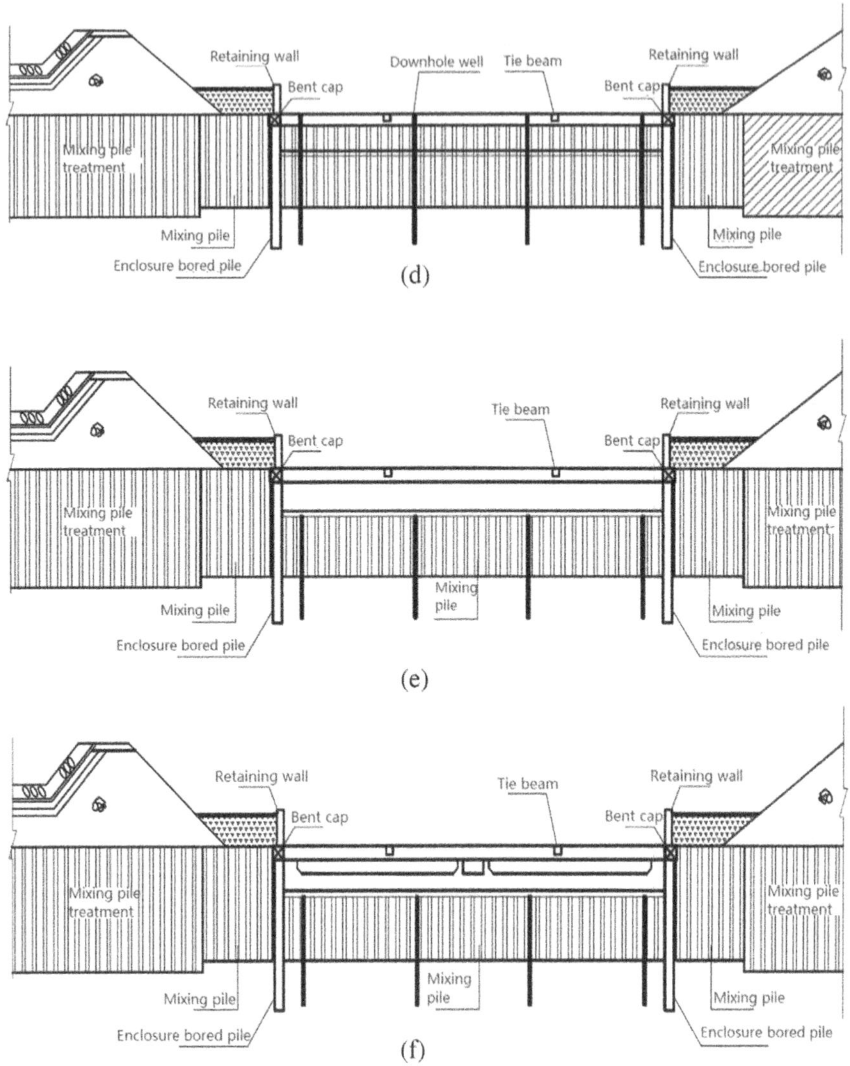


**Fig. 5.27** (continued)

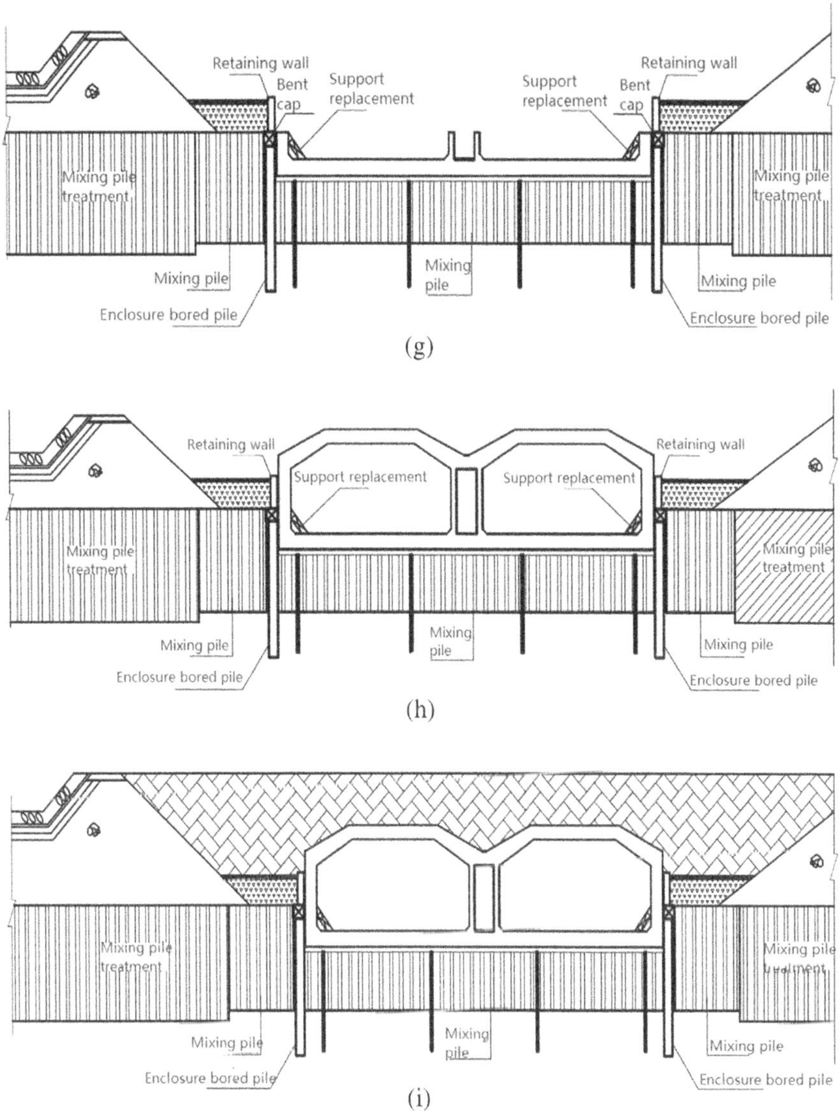


**Fig. 5.27** (continued)

parallel footage is 1–2 m each time. It is necessary to follow the early strong support in time and bear all loads to ensure the stable equilibrium of the process of the construction of the tunnel structure.

(2) When the face of the tunnel face is stable during the construction of the tunnel, the short steps should be used to reserve core soil to approximate the full cross-section construction method, as shown in Fig. 5.28c. Tunnels with class III, IV

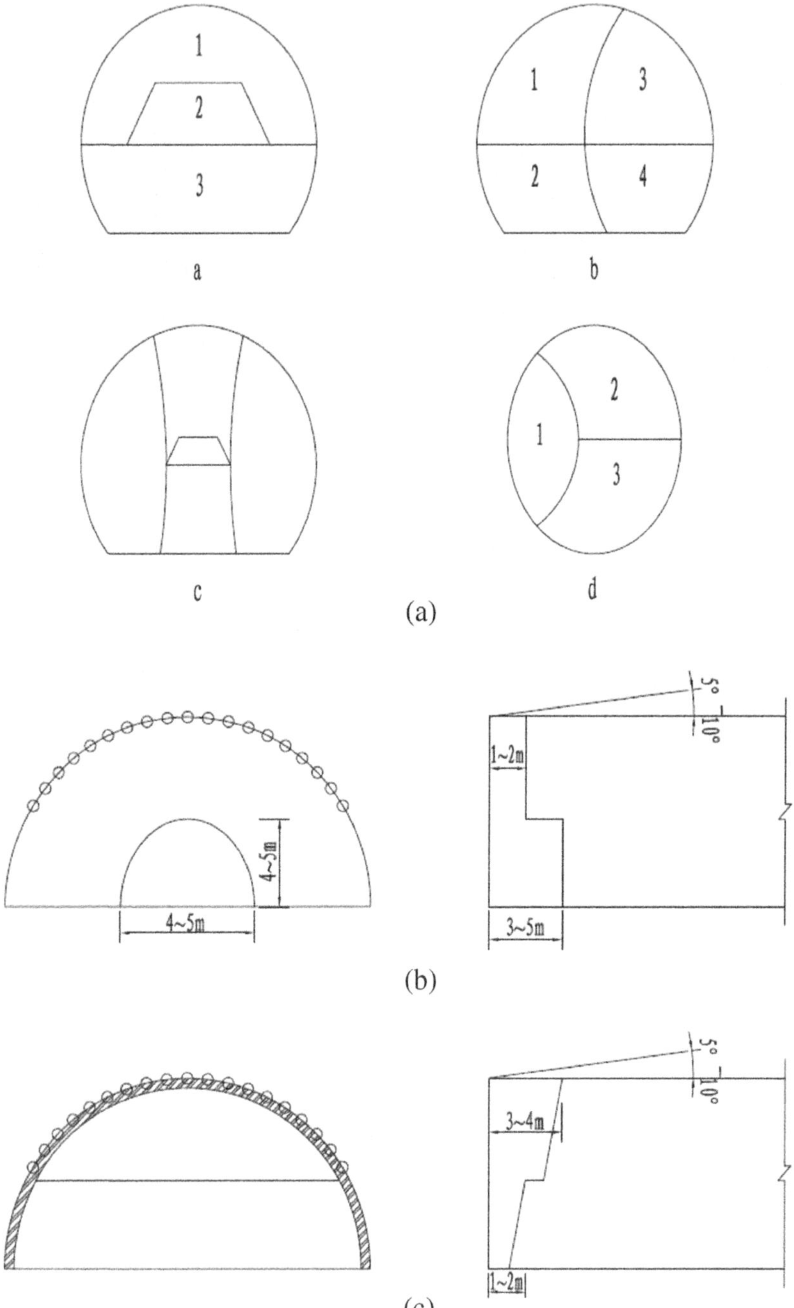


**Fig. 5.28** **a** Diagram of various methods for the partial excavation and support mentioned in tunnel planning or manuals; **b** Diagram of the method of construction of the approximate full cross-section with a moderate advancement of the lower guide; **c** Diagram of the approximate full cross-section method of construction of the short step reserved core soil

and V surrounding rock should adopt short steps and reserve core soil for the upper and lower steps under a strong pre-support. The upper and lower steps are parallel to each other by 1–2 m. They should follow the strong initial support in time and bear all loads to ensure the stable equilibrium of the process of the construction of the tunnel structure.

**Problem 5.2.4**

It is about the design and method of construction of crack control for the deep-buried circular pipe culvert in mountainous areas. With the implementation of the new national regional coordinated development plan, many mountain roads need to be built. Setting up deep-buried circular pipe culverts is an economic and practical method for high-fill mountain roads, but some deep-buried circular pipe culverts are cracked. Therefore, it is necessary to study and improve the design and method of construction of crack control for the deep-buried circular pipe culvert in mountainous areas from the perspective of mechanical analysis.

**Solution** In view of the existing code for deep-buried circular pipe culverts, which does not explain how to avoid the risk of cracking from the mechanical concept, the mechanical method is directly used to analyze how to avoid the risk of cracking for deep-buried circular pipe culverts, and to find the method of control of cracking for deep-buried circular pipe culverts.

The first step is to analyze the cause of the cracking of deep-buried circular pipe culverts directly by the mechanical method;

From Fig. 5.29a, it can be seen that the design load of deep-buried circular pipe culverts

$$P_{design} = H \times D \tag{5.2.4}$$

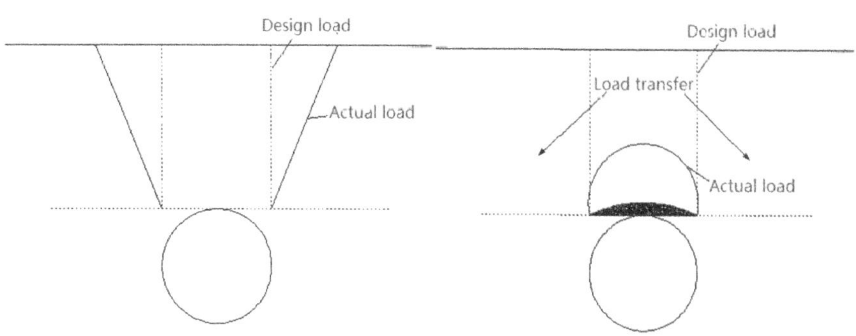


A) The load of the circular pipe culvert is similar to an inverted trapezoid distribution; b) The load of the circular pipe culvert is similar to the collapse distribution

**Fig. 5.29** Diagram of the stress characteristics of deep-buried circular pipe culverts

Because the rigidity of the circular pipe culvert is greater than that of the subgrade, the design load of the circular pipe culvert is a rectangular distribution, which is smaller than the actual load of the circular pipe culvert and is similar to the inverted trapezoid distribution. The excessive stress of shallow-buried circular pipe culverts is within the safe range; the excessive stress of deep-buried circular pipe culverts will lead to the cracking of the circular pipe culvert. In fact, the weak part of or link in the deep-buried circular culvert is the cracking of the circular culvert. The transfer or concentration of the induced adverse force and energy to the circular pipe culvert structure should be prevented. That is, the actual load of the deep-buried circular pipe culvert

$$\mathrm{P_{actual}} = \mathrm{H} \times (\mathrm{D} + \mathrm{H}\sin\alpha) \tag{5.2.5}$$

where H is the filling height above the top of the circular pipe culvert; D is the diameter of the circular pipe culvert; α is the actual load expansion angle; $\mathrm{P_{design}}$ is the design load of the deep-buried circular pipe culvert; $\mathrm{P_{actual}}$ is the actual load of the deep-buried circular pipe culvert.

$$\text{Obviously, } \mathrm{P_{actual}} > \mathrm{P_{design}} \tag{5.2.6}$$

If $\mathrm{P_{actual}}$ exceeds the allowable stress of the deep-buried circular pipe culvert, it will inevitably produce cracks.

In the second step, the mechanical method is directly used to find the method of load reduction for the deep-buried circular pipe culvert;

According to Figs. 5.29b and 5.30, we can see that at the top of the deep-buried circular pipe culvert, there are elastic soil arch cushions, such as thick D/6–D/4 foam and other elastic-plastic materials, and then a filling construction. The subgrade block or particle has a sinking tendency on the cushion, but the cushion is obstructing. The subgrade has formed the soil arch, and the load on the upper part of the soil arch has been transferred to the nearby subgrade. The equilibrium formula of this is as follows:

$$P_1 \cos\alpha_1 + P_2 \cos\alpha_2 + T = W \tag{5.2.7}$$

where $\mathrm{P_1}$, $\mathrm{P_2}$—mutual supporting force between subgrade blocks or particles; W—subgrade gravity; T—subgrade pressure at the top of circular pipe culvert (subgrade pressure T is as small as possible).

Therefore, the design load of the circular pipe culvert is more rectangular than the actual load of the circular pipe culvert, which is similar to the collapse distribution. The upper load of the circular pipe culvert is transferred to the nearby subgrade to bear, so that either the shallow or the deep-buried circular pipe culvert will be in a safe range, and the circular pipe culvert will not crack. That is, the actual load of the deep-buried circular pipe culvert

$$\mathrm{P_{actual}} = \mathrm{D}/2 \times \mathrm{D} \tag{5.2.8}$$

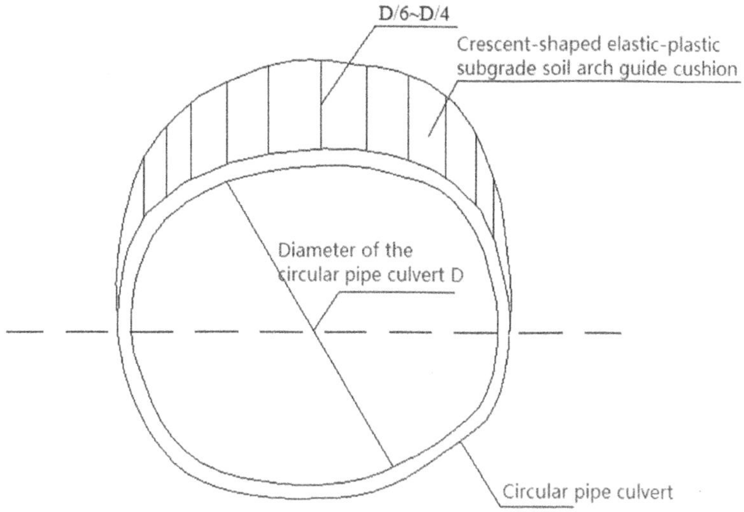


**Fig. 5.30** Diagram of the forming of the soil arch guide cushion on the top subgrade of deep-buried circular pipe culverts

where H is about equal to D/2, far less than the filling height above the top of the circular pipe culvert;

$$\text{Obviously, } P_{\text{actual}} < P_{\text{design}} \tag{5.2.9}$$

In this way, part of the filling load is transferred to the nearby subgrade to bear, and the deep-buried circular pipe culvert will not naturally produce cracks.

**Problem 5.2.5**
Many highway tunnels in mountainous areas built in the 1980–1990s in China adopted the scheme of shortening the tunnel length and formed a cutting slope at the entrance of the tunnel to achieve the purpose of saving funds. However, the risk of slope collapse at the tunnel portal as shown in Fig. 5.31 was also created.

The side slope with a height of 1–1.5 m on the top of a tunnel portal is cut out by sliding blocks. The upper cut is about 1 m wide and 2–3 m deep, and the lower cut is protruding and leaking. The sliding block is about 15 m long and about 10 m wide and 6–10 m wide, with a volume of about 500–1,000 $m^3$. The flow of the tunnel is about 17,000 vehicles/day, exceeding the design flow of 15,000 vehicles/day. The sliding block on the upper part of the tunnel portal has seriously threatened the safety of traffic and personnel. How can this be dealt with? It is necessary to study the prevention and control methods of the slope collapse risk at the entrance of the highway tunnel in mountainous areas from the perspective of a mechanical analysis.

**Solution** In view of the advantages and disadvantages of the existing prevention and control methods for the landslide risk at the entrance of highway tunnels in

**Fig. 5.31** Photo of the risk of block collapse at the tunnel portal slope

mountainous areas, the paper analyzes the landslide risk at the entrance of highway tunnels in mountainous areas directly from the perspective of safety management and the use of mechanical methods to find more safe and stable methods of prevention and control.

Temporary treatment suggestions: (1) Close the tunnel in the front and back nodes of the tunnel and organize two-way traffic in adjacent tunnels; organize vehicles to bypass along the road intersection; (2) Close the upper cut of the sliding block; (3) Organize an assessment of the risk of the sliding block and implement the reinforcement and reconstruction scheme.

Evaluation of the four reinforcement and reconstruction schemes: (1) The scheme of anchor bolt strengthening the sliding block: considering that the lower part of the sliding block has been cut out, the sliding block may be triggered by the drilling impact of the down-hole drill, and there are construction safety risks, and the anchoring effect is not good; moreover, it is difficult to guarantee the quality of the anchoring project after reinforcement, so the scheme is not appropriate; (2) Anti-slide pile supporting scheme: Considering that the lower cut of the slide block is only 1–1.5 m away from the lower slope of the portal, there is no equilibrium in front of the anti-slide pile, the anti-slide pile foundation pit blasting construction may trigger the sliding of the slide block, so there are construction safety risks, and it is difficult to put into place a guarantee of an anti-slide effect, so the scheme is not appropriate; (3) Excavation of a sliding block unloading and reinforcement scheme: the scheme may affect the system of equilibrium at the top of adjacent tunnels, increase the stress on the tunnel lining, and may cause the cracking in the tunnel lining; and the subsequent stability of the portal slope; in addition, the roads in mountainous areas are very narrow, the re-construction of roads, waste disposal areas and other links affect the ecological environment, which does not meet the requirements of a beautiful

China; (4) Extension of the open cut tunnel: first remove the dangerous rocks on the side slope. Considering that the sliding block incision is 1–1.5 m above the top of the tunnel portal, and the fill on the top of the open cut tunnel is more than one time the tunnel diameter height after the completion of the open cut tunnel, the formula for the equilibrium of the backfill earthwork on the top of the open cut tunnel is as follows:

$$P_1 \cos \alpha_1 + P_2 \cos \alpha_2 + T = W \tag{5.2.7}$$

where $P_1$, $P_2$—mutual support force between filling blocks or particles; W—filling gravity; T—filling pressure at the top of the open cut tunnel (filling pressure T is as small as possible).

It is a good thing to excavate some sliding blocks and fill them symmetrically on both sides and on the top of the open cut tunnel, which is not only the waste slag filling site but also the equilibrium of the remaining sliding blocks, without affecting the ecological environment. Check the stability of the upper slope of the extended open cut tunnel backfill, and then reinforce it with anchor bolts or anchor rods, as shown in Fig. 5.32.

**Problem 5.2.6**

There are many areas in coastal cities which belong to the soft soil stratum. With the modernization of the city, it is not suitable to build elevated expressways on the upper part of a main urban road in the existing high-grade business district and high-grade residential district, so we have to build underground expressways. In this way, when the strength of the soil layer under an urban main road is good, the process of

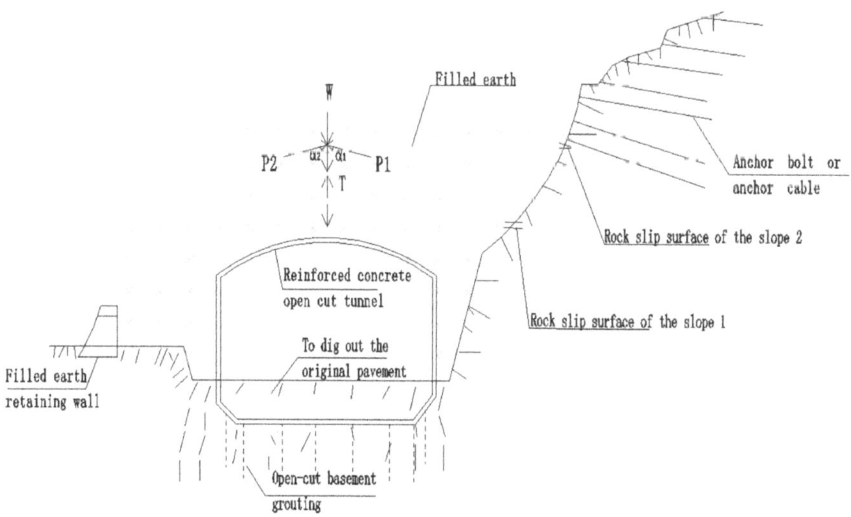


**Fig. 5.32** Diagram of the extended open cut tunnel

the underground expressway and metro shield interval is generally to excavate and pour the underground expressway tunnel first, and then construct the metro interval. At this time, no pre-reinforcement and settlement reduction measures are required, as shown in Fig. 5.33a. However, if the construction area of the project is in the condition of muddy soft soil, the sinking of the underground expressway tunnel is quite obvious, and the structure of the metro shield interval is also unstable in muddy soft soil. At this time, if no measures for pre-reinforcement and settlement reduction are taken, the feasibility of the project cannot be satisfactory, so it is necessary to add some necessary and feasible procedures in the above construction steps to meet the operability of the project. Based on the principle of super front support in mountain tunnels, some measures for pre-reinforcement and settlement reduction must be added on the basis of the original process of engineering construction to find an effective solution for the construction of an underground expressway tunnel and metro shield interval in muddy soft soil.

**Solution** In view of the problems existing in the construction of an underground expressway tunnel and metro shield section in the muddy soft soil under an urban main road. In other words, the construction is not safe, the frame structure of the tunnel of the underground expressway and the lining structure of the shield segment of the subway are not stable. Some necessary pre-reinforcement measures must be adopted, and established models need to be simplified based on the Protodyakonov's Pressure Arch Theory to analyze and estimate the minimum thickness of the upper covering layer of the shield segment lining structure.

(1) Before excavation of a foundation pit, the retaining structure (diaphragm wall) should be constructed first. After the completion of the construction of the retaining structure, the three-axis cement mixing pile is used for foundation reinforcement, and the mucky soil above the bottom plate of the underground expressway tunnel should be weakly reinforced to facilitate the excavation of the foundation pit. In contrast, the mucky soil under the bottom plate of the underground expressway tunnel should be strongly reinforced, which can not only ensure the stability of the driving of the subway shield and the stability deformation of the shield segment lining structure, but it can also limit the sinking of the underground expressway tunnel structure.
(2) According to the design, uplifting piles and column piles (also as uplifting piles) are to be constructed. The first function of the uplifting pile is to support the frame structure of the upper underground expressway tunnel and control the settlement deformation; the second is that when the shield is pushed forward, the upper soil mass will be uplifted and deformed, and the frame structure of the underground expressway tunnel may float upwards, and the uplifting pile plays an anti-floating role at this time.
(3) The shield crossing conditions should be reserved between the column piles, so that there is enough space for the later shield crossing, so as not to affect the stability of the pile foundation. According to experience, the clear distance d between the anti-pulling and settlement reducing pile, the column pile and the shield structure should be at least $\geq 1.5$ m.

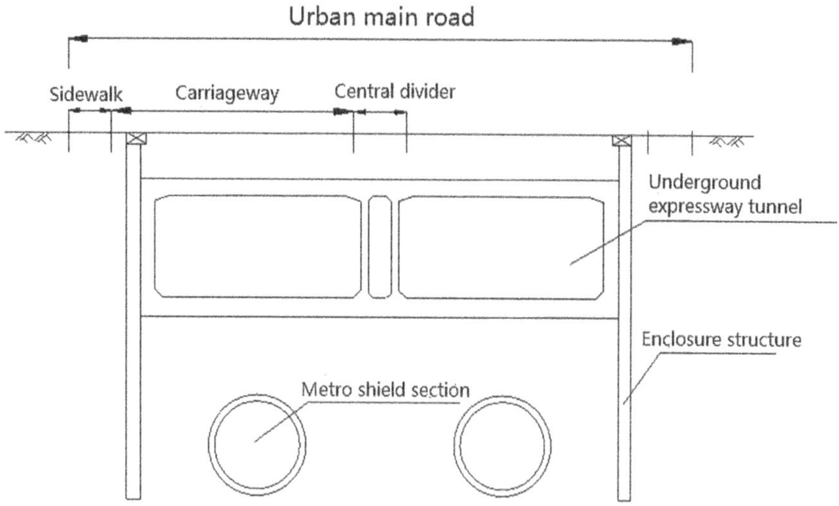


(a)

**Fig. 5.33** **a** Diagram of the separate mode of construction of the underground expressway tunnel and metro shield section; **b** Working condition 1 of the separate construction of the underground expressway tunnel and metro shield section: diagram for the construction of the diaphragm wall, the reinforcement of mixing piles and the construction of uplifting piles (including column piles); **c** Working condition 2 of the separate construction of the underground expressway tunnel and metro shield section: diagram of excavating soil to the bottom of the first support of a foundation pit and pouring the crown beam and the first concrete support; **d** Working condition 3 of the separate construction of the underground expressway tunnel and metro shield section: diagram of excavating soil to the bottom of the second support, setting the second steel support and applying pre-stress; **e** Working condition 4 of the separate construction of the underground expressway tunnel and metro shield section: diagram of excavating soil to the bottom of the third support, setting the third steel support and applying pre-stress; **f** Working condition 5 of the separate construction of the underground expressway tunnel and metro shield section: diagram of excavating soil to the pit bottom, pouring of a plain concrete cushion and the frame bottom plate of the underground expressway tunnel; **g** Working condition 6 of the separate construction of the underground expressway tunnel and metro shield section: diagram of removing the third steel support, pouring the side wall, and adding support after the side wall reaches the design strength; **h** Working condition 7 of the separate construction of the underground expressway tunnel and metro shield section: diagram of removing the second steel support and pouring the top plate of the frame; **i** Working condition 8 of the separate construction of the underground expressway tunnel and metro shield section: diagram of removing the first support and the support in the frame after the top plate reaches the design strength, and dismantling the column; **j** Working condition 9 of the separate construction of the underground expressway tunnel and metro shield section: diagram of constructing the additional waterproof layer of the roof, filling and covering the soil, restoring the surface of the urban main road, and conducting subsequent shield tunneling (In the figure, h is the thickness of overburden, d is the clear distance between the anti-pulling and settlement reducing pile, column pile and shield structure); **k** Diagram of a simplified calculation model of the estimation of the overburden thickness based on the Protodyakonov's Pressure Arch Theory

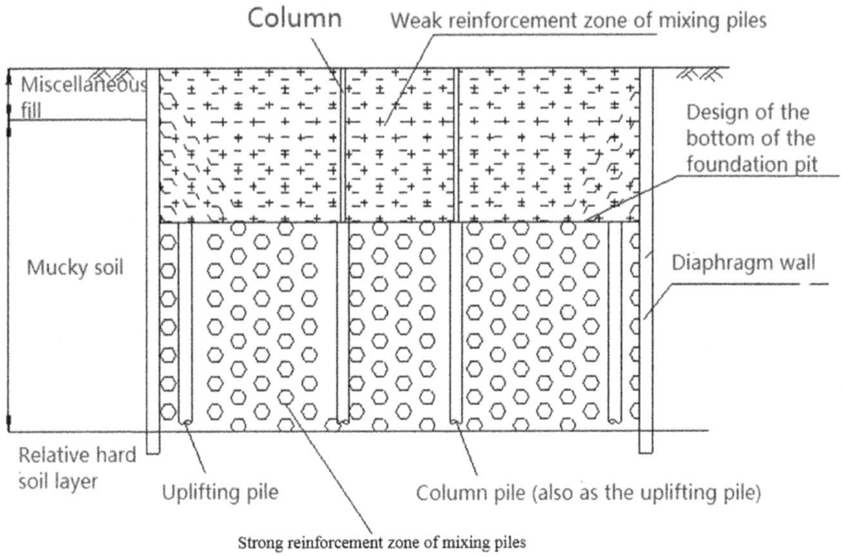


(b)

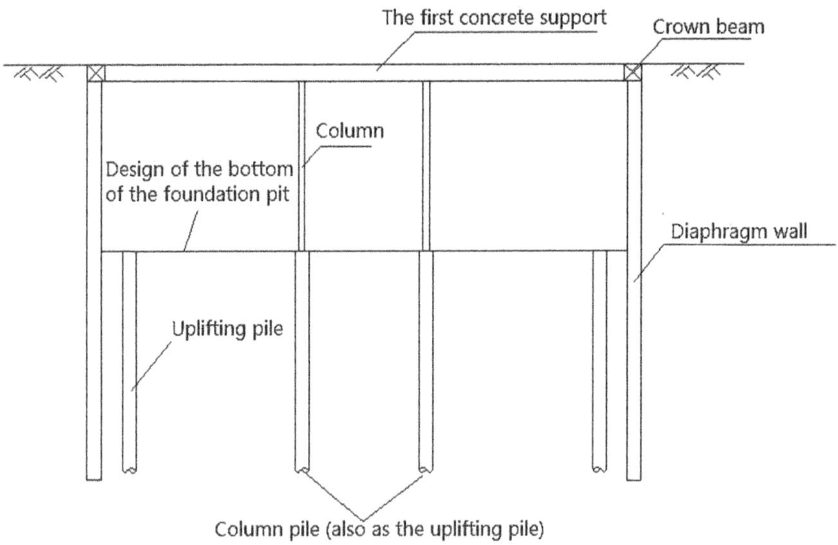


(c)

**Fig. 5.33** (continued)

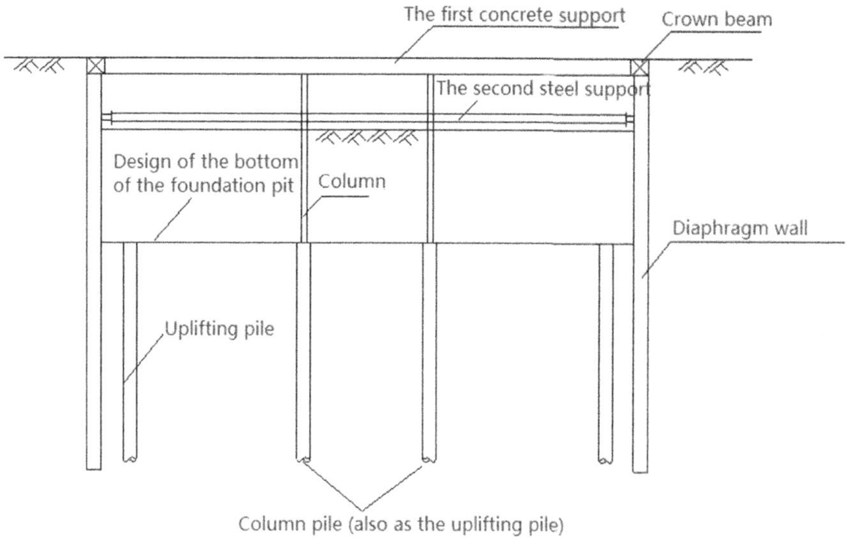


(d)

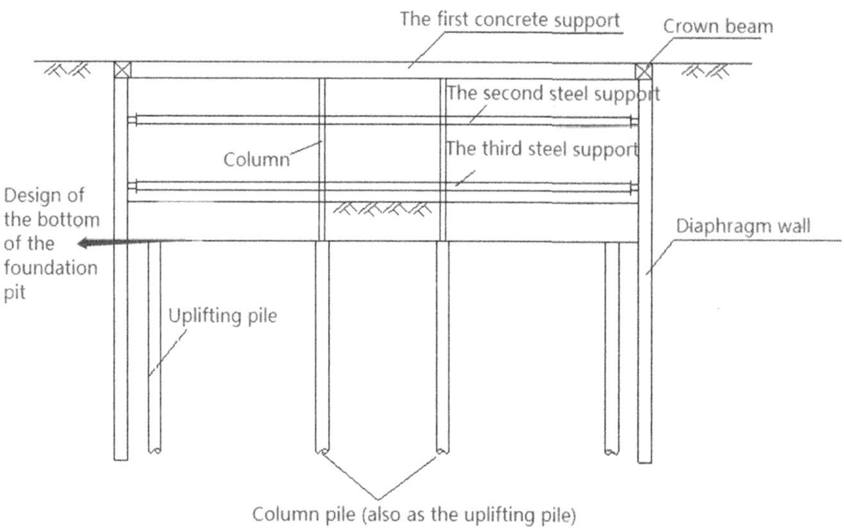


(e)

**Fig. 5.33** (continued)

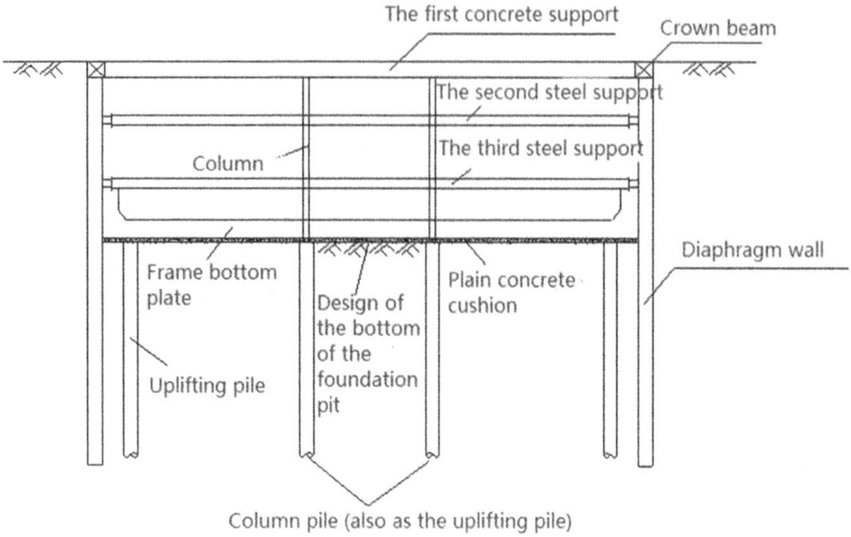


(f)

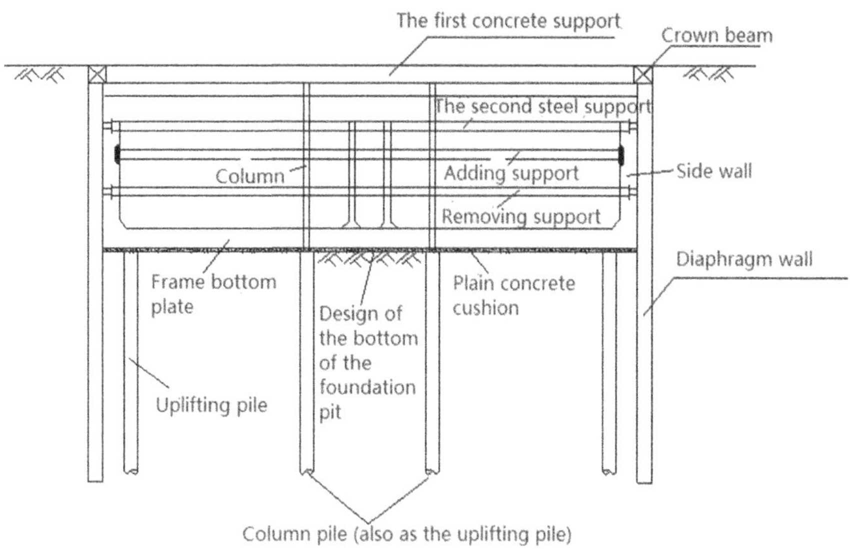


(g)

**Fig. 5.33** (continued)

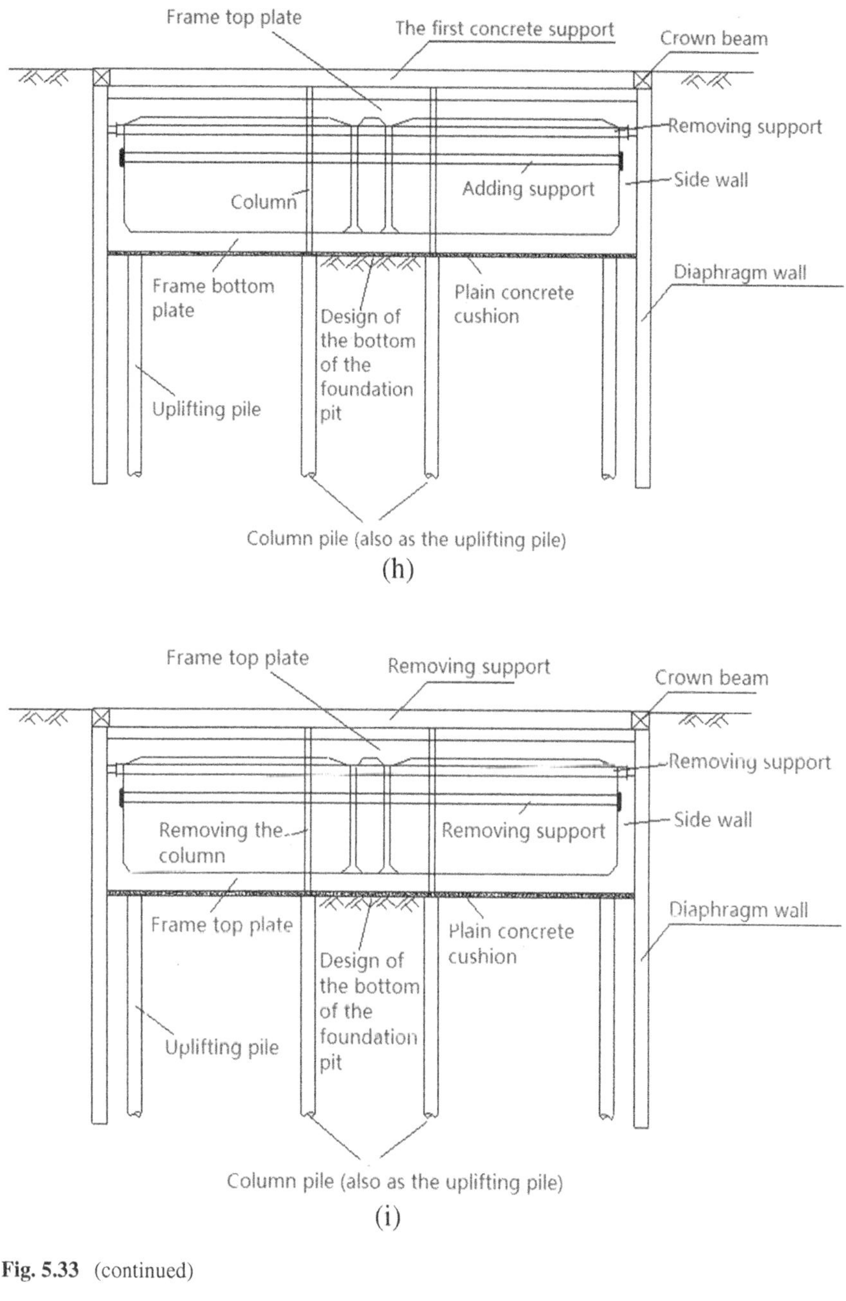


**Fig. 5.33** (continued)

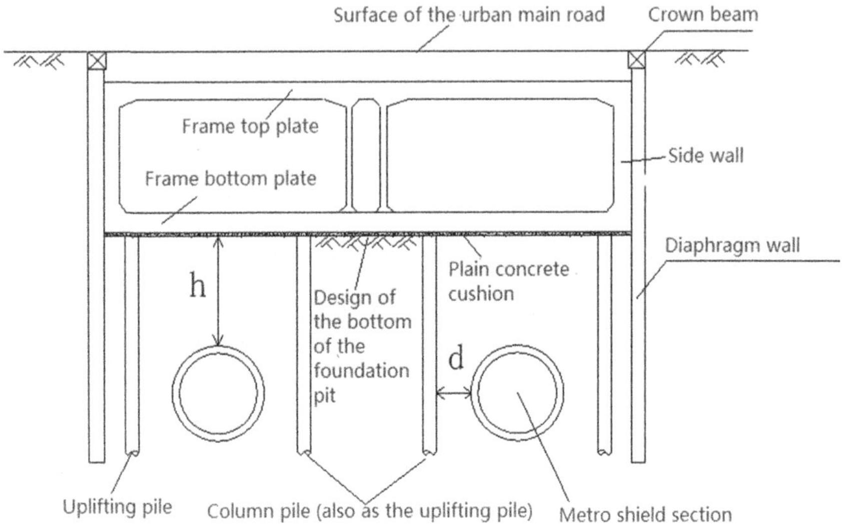


(j)

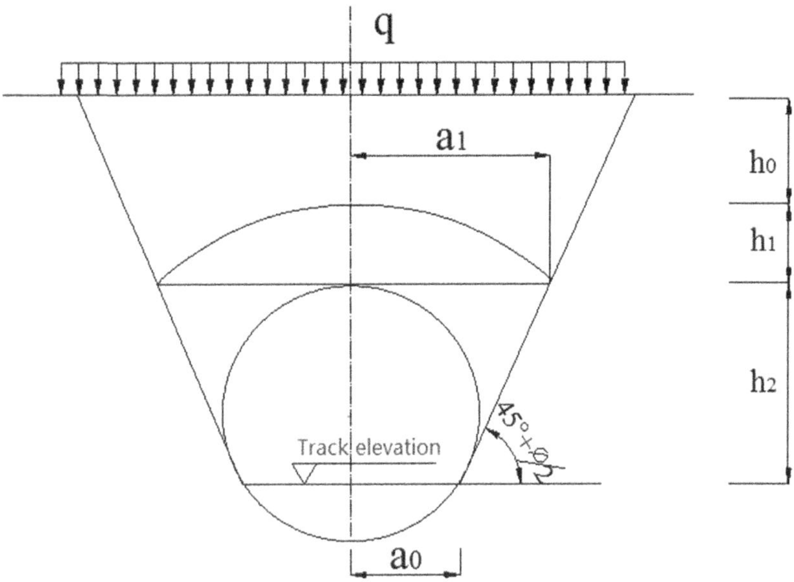


(k)

**Fig. 5.33** (continued)

(4) In terms of the thickness of the upper overburden of the shield interval lining structure, a simplified model has been established for analysis and estimation based on the Protodyakonov's Pressure Arch Theory. The model is shown in Fig. 5.33k. The rock mechanics parameters can be obtained from the in situ sampling of the silt soil after strong reinforcement by the indoor testing. The minimum cover thickness h is calculated by the following formula:

$$b_1 h^2 + b_2 h + b_3 \leq 0$$

where $b_1$, $b_2$, $b_3$ are coefficients, respectively calculated as follows:

$$b_1 = \gamma \frac{\tan^2(45^\circ - \varphi/2)}{R_c} F$$

$$b_2 = \gamma a_1 \frac{2\tan(45^\circ - \varphi/2)}{R_c} F + q \frac{2\tan^2(45^\circ - \varphi/2)}{R_c} F - 1$$

$$b_3 = \left( q a_1 - \frac{2a_1^2}{3f} \gamma \right) \frac{2\tan(45^\circ - \varphi/2)}{R_c} F + \frac{a_1}{f}$$

In the formula, h—thickness of the overburden, $h = h_0 + h_1$;
$\gamma$ and $\varphi$ are rock gravity and the internal friction angle, where the silt soil after strong reinforcement can be sampled and obtained by indoor testing;
$R_c$—The uniaxial ultimate compressive strength of the rock can be obtained by means of a laboratory test by sampling the silt soil after strong reinforcement;
$a_1$—Pressure arch half span, $a_1 = a_0 + h_2 \tan(45^\circ - \varphi/2)$, $a_0$ are half of the rail surface span;
$h_2$—The distance between the rail surface and the top of the shield segment;
q—The total load of the covering soil on the top of the tunnel frame and on the tunnel frame;
$F$—Safety factor, which should generally be adopted in the project $F = 8$;
$f$—Protodyakonov coefficient.

**Problem 5.2.7**

Many areas in coastal cities belong to the soft soil stratum. With the modernization of the city, it is not suitable to build an elevated expressway on the upper part of the main urban road in the existing high-grade business district and high-grade residential district, so we have to build an underground expressway. In this way, when the strength of the soil layer under the urban main road is good, the process of the underground expressway and the rail crossing section is generally to construct the retaining structure first, excavate the foundation pit, and then pour the rail crossing section structure and the structure for the frame of the underground expressway tunnel from the bottom to the top in turn. At this time, no pre-reinforcement and settlement reduction measures are required, as shown in Fig. 5.34a. However, if the

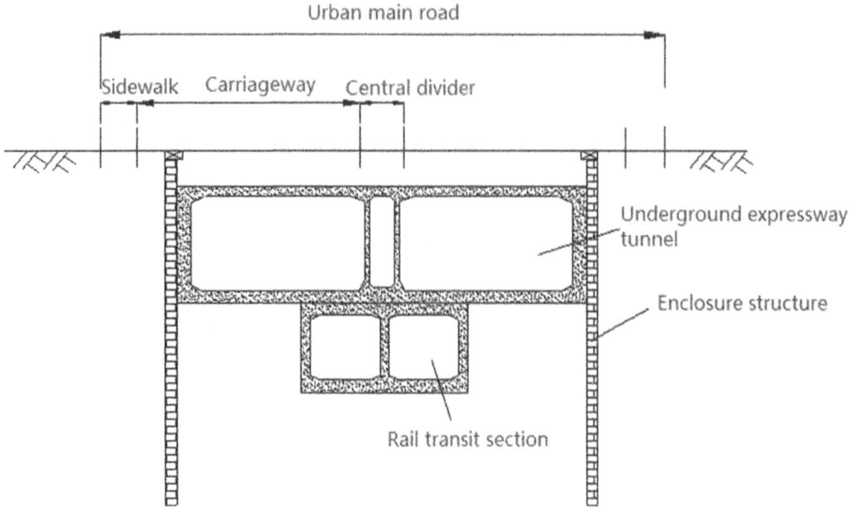


(a)

**Fig. 5.34** **a** Diagram of the joint construction mode of the underground expressway tunnel and the rail transit section under the urban main road; **b** Diagram of the joint construction mode of the underground expressway tunnel and the rail transit section under the urban main road after the local optimization design; **c** Working condition 1 of the joint construction of the underground expressway tunnel and the rail transit section: diagram for the construction of the diaphragm wall (embedded inclinometer probe), reinforcement of the cement mixing piles and construction of the settlement reducing piles (including column piles); **d** Working condition 2 of the joint construction of the underground expressway tunnel and the rail transit section: diagram of excavating the soil to the bottom of the first support of a foundation pit and pouring the crown beam and the first concrete support; **e** Working condition 3 of the joint construction of the underground expressway tunnel and the rail transit section: diagram of excavating the soil to the bottom of the second support, setting the second steel support and applying pre-stress; **f** Working condition 4 of the joint construction of the underground expressway tunnel and the rail transit section: diagram of excavating the soil to the bottom of the third support, setting the third steel support and applying pre-stress; **g** Working condition 5 of the joint construction of the underground expressway tunnel and the rail transit section: diagram of excavating the soil to the bottom of the fourth support, setting the fourth concrete support and applying pre-stress; **h** Working condition 6 of the joint construction of the underground expressway tunnel and the rail transit section: diagram of excavating the soil to the bottom of a large foundation pit, then excavating to the bottom of a small foundation pit according to the design, and pouring the plain concrete cushion; **i** Working condition 7 of the joint construction of the underground expressway tunnel and the rail transit section: from the bottom to the top, pouring the bottom plate and the side wall of the structure of the rail transit section in turn, backfilling at both sides, and pouring the bottom plate of the frame of the underground expressway tunnel; **j** Working condition 8 of the joint construction of the underground expressway tunnel and the rail transit section: diagram of removing the fourth support, pouring the side wall of the tunnel frame and replacing the support; **k** Working condition 9 of the joint construction of the underground expressway tunnel and the rail transit section: diagram of removing the third support and pouring the top plate of the frame of the underground expressway tunnel; **l** Working condition 10 of the joint construction of the underground expressway tunnel and the rail transit section: diagram of removing the second support and the first support in sequence upward, and then removing the replaced support; (m) Working condition 11 of the joint construction of the underground expressway tunnel and the rail transit section: diagram of cutting off the lattice column, constructing the additional waterproof layer on the top plate of the tunnel frame, conducting soil backfill, and restoring the surface of the urban main road

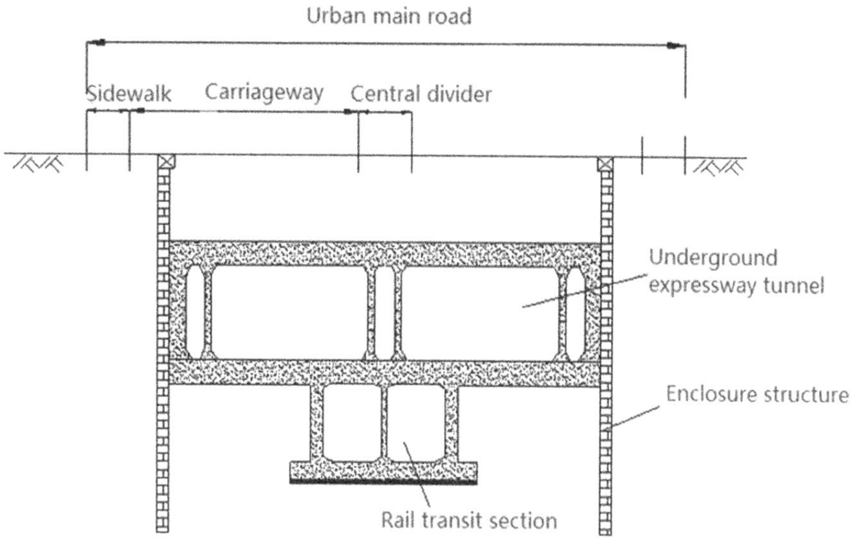


(b)

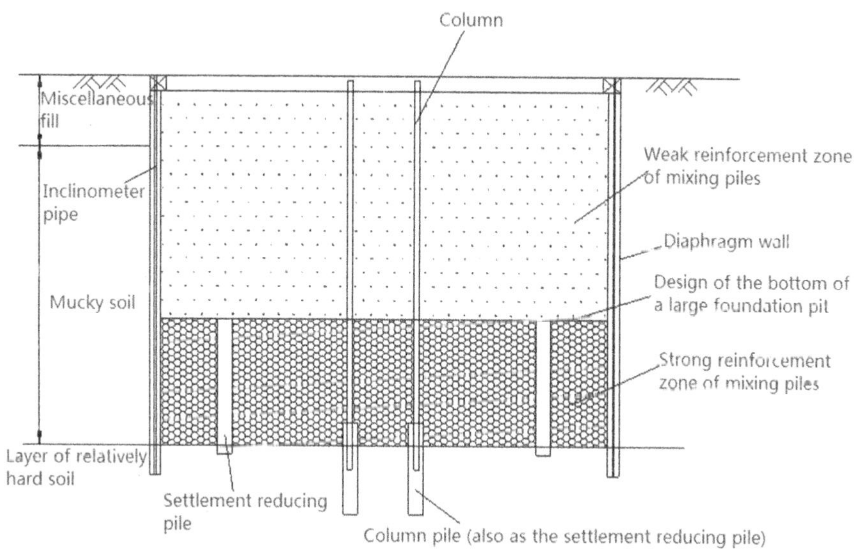


(c)

**Fig. 5.34** (continued)

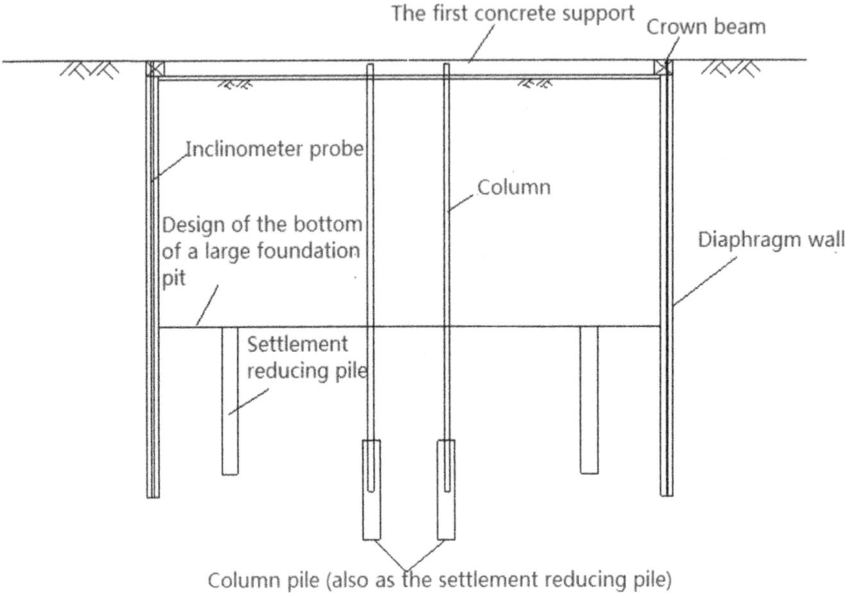


(d)

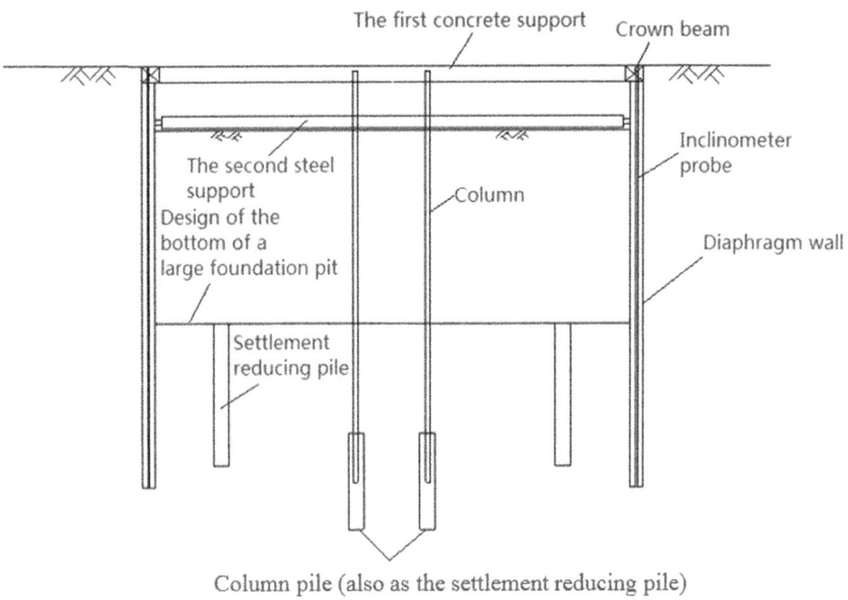


(e)

**Fig. 5.34** (continued)

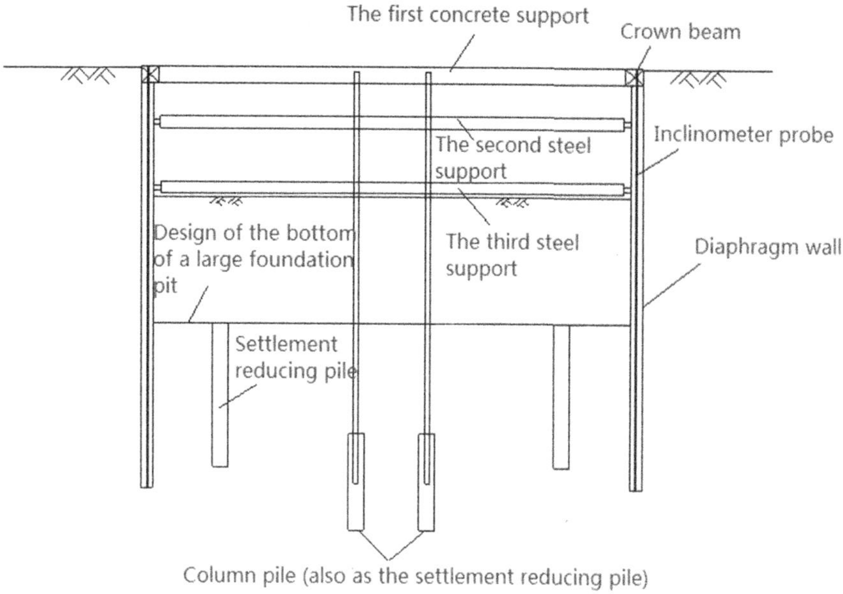


(f)

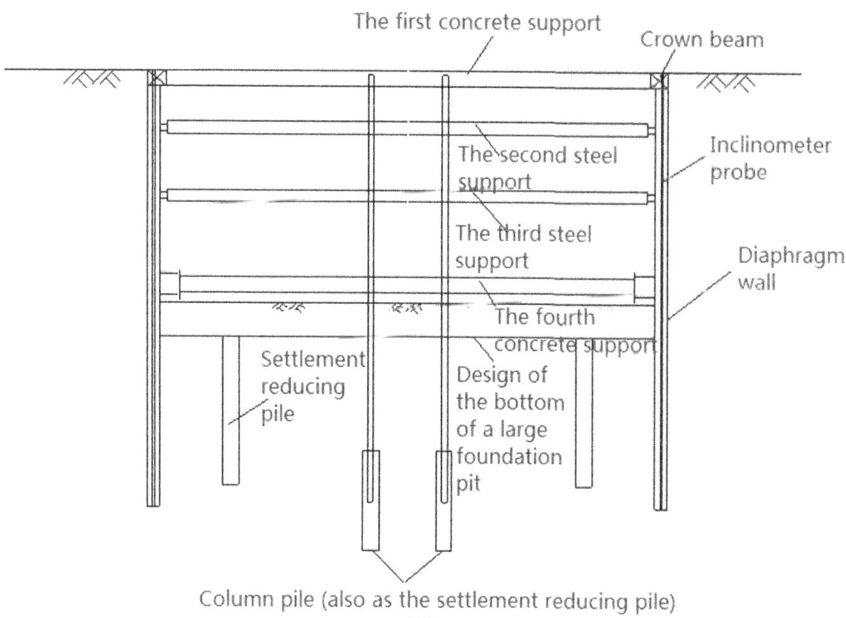


(g)

**Fig. 5.34** (continued)

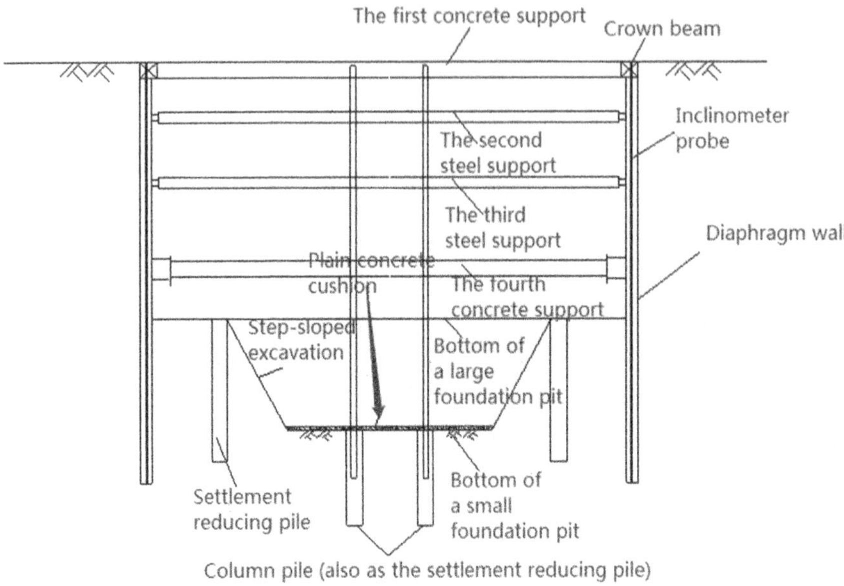


(h)

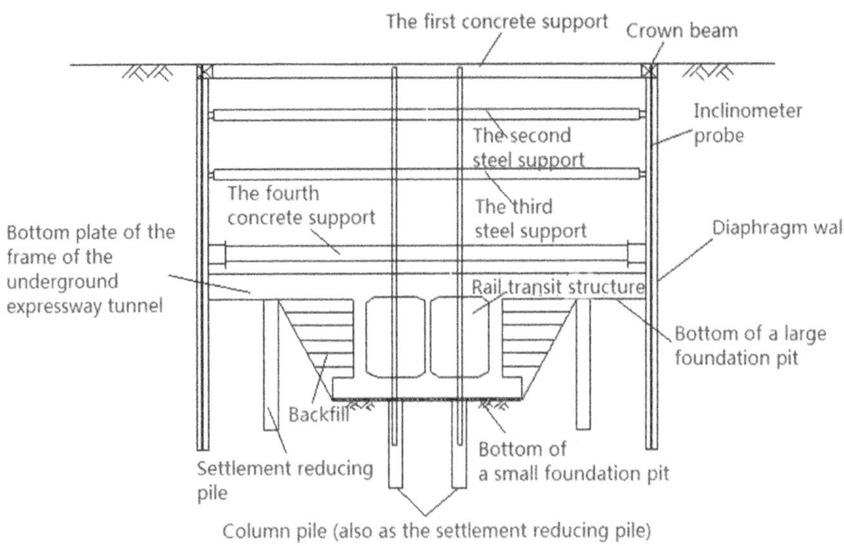


(i)

**Fig. 5.34** (continued)

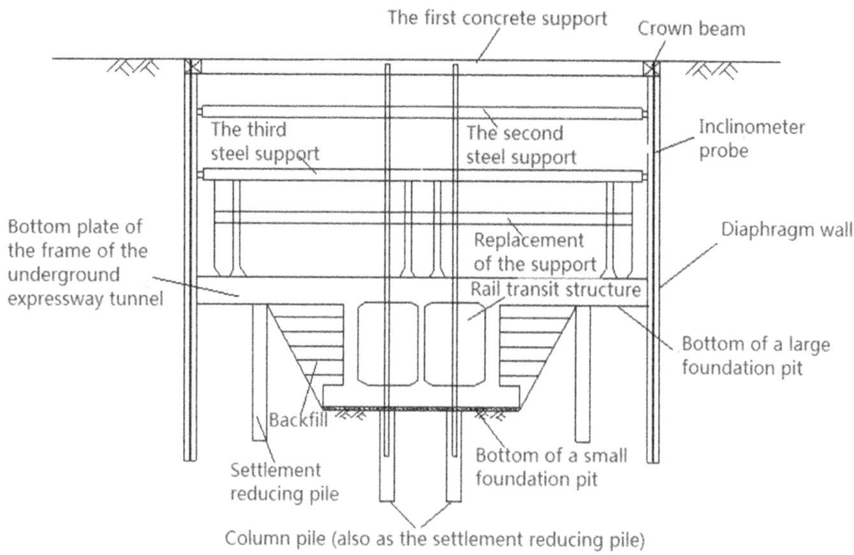


(j)

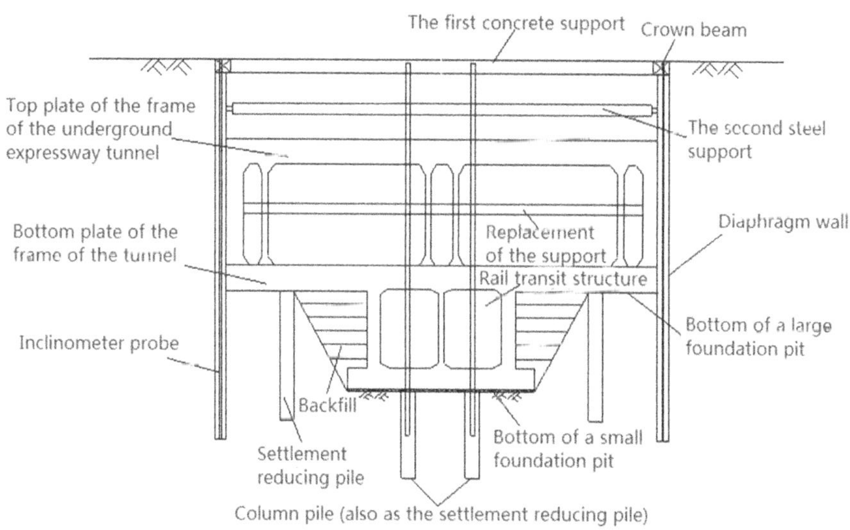


(k)

**Fig. 5.34** (continued)

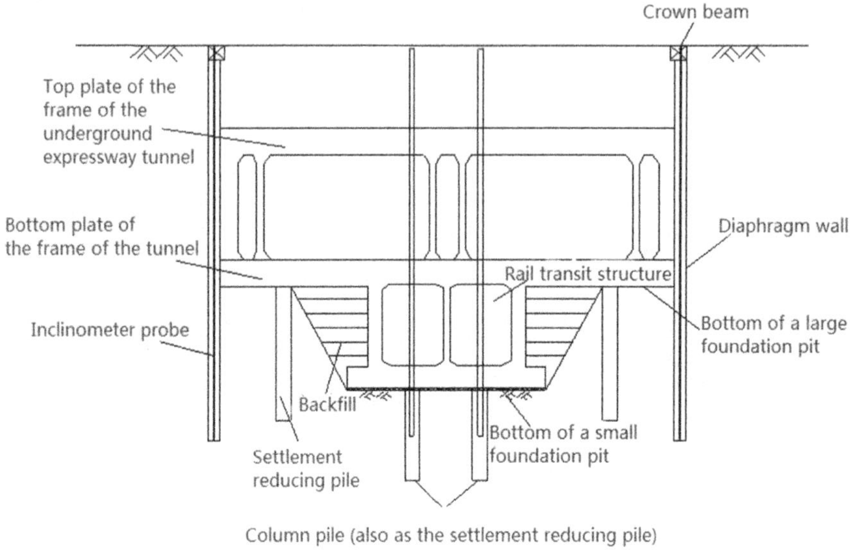


(l)

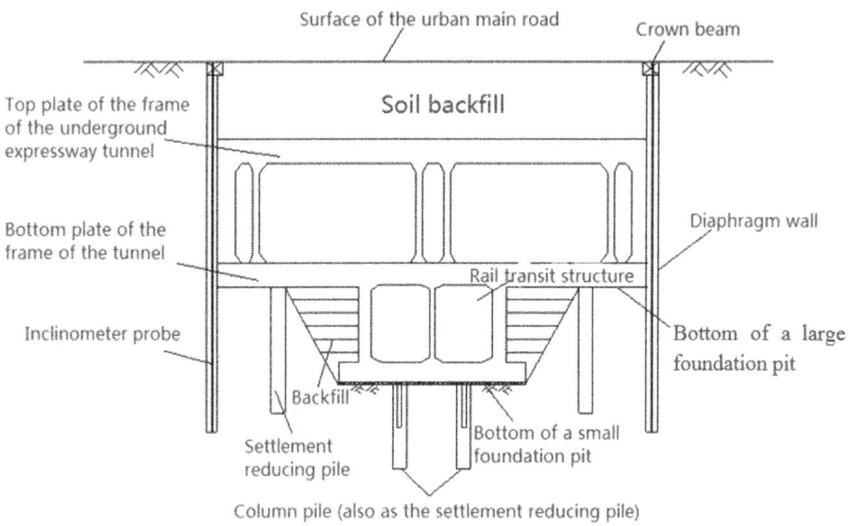


(m)

**Fig. 5.34** (continued)

soil in the construction area of the project is muddy and soft, the sinking of the underground expressway tunnel would be quite obvious, and the structure of the rail transit section would also be unstable in muddy soft soil. At this time, if no pre-reinforcement and settlement reduction measures are taken, the feasibility of the project would not be satisfactory, so it would be required to add some necessary and feasible procedures in the above construction steps to meet the operability of the project. Based on the principle of a super front support in mountain tunnels, some necessary pre-reinforcement and settlement reduction measures would be added on the basis of the original construction process. In addition, some local optimization design adjustments would also be made on two kinds of underground structures to find an effective solution for the construction of the underground expressway tunnel and rail crossing section in muddy soft soil.

**Solution** In view of the problems existing in the joint construction of the underground expressway tunnel and the rail transit section in the muddy soft soil under the urban main road, some necessary and feasible pre-reinforcement measures would be taken and some optimal designing would be done.

(1) Before excavation of a foundation pit, the retaining structure (diaphragm wall) should be constructed first. After the completion of the construction of the retaining structure, the three-axis cement mixing pile is used for the foundation reinforcement, and the silt soft soil above the bottom plate of the frame of the underground expressway tunnel is used for the weak reinforcement to facilitate the excavation of a large foundation pit. The cement content in the weak reinforcement area of the cement mixing pile is 7–9%.
(2) In the excavation of a foundation pit, the lower the excavation, the greater the construction risk, especially in the "pit-in-pit" of muddy soft soil, so there will be construction risks (the first risk of which is, at the bottom plate of the frame of the underground expressway tunnel. The second risk is that the bottom of the large foundation pit is designed to continue to be excavated downward to the bottom plate of the rail crossing structure at the bottom of the small foundation pit). The following design and construction considerations are made:

2.1 The cement mixing pile is stronger than the upper soil mass in the bottom of the large foundation pit. The mud soil under the bottom plate of the frame of the underground expressway tunnel is strengthened. The cement content in the strong reinforcement area is 15–20%, which can not only ensure the stability of the structure of the construction track section, but it can also ensure the stability of the later operations on the track section;
2.2 Continue to excavate downward from the bottom of the designed large foundation pit to the bottom of the track crossing structure of the small foundation pit, and the slope excavation can also reduce the construction risk of "pit-in-pit" to a certain extent;
2.3 In order to control the settlement and deformation of the tunnel structure and the rail crossing structure of the underground expressway, 10–15 m of

settlement reducing piles and 25–40 m of column piles (also as settlement reducing piles) should be constructed.

(3) The frame structure of the underground expressway tunnel and the structure of the rail transit section are generally designed as separate structures, which are independent individuals, which is not conducive to controlling differential settlement, as shown in Fig. 5.34a. The frame structure of the highway tunnel is integrated with the rail crossing structure (as shown in Fig. 5.34b), which is more conducive to deformation coordination and to the sharing of the upper load together, just like the raft foundation (the most common is the pedestrian bamboo raft in the river) has stronger overall stability; moreover, the two types of structures are arranged with settlement reducing piles under the bottom plate, which is more conducive to settlement control. The co-construction of the two structures can save on the construction period and on the project cost under the condition of ensuring safety.

(4) In order to reduce the additional load on the upper part of the bottom plate of the rail transit structure and the settlement deformation of the upper structure, the area (width) of the bottom plate of the rail transit structure can be expanded, which is more conducive to stress diffusion, and then the settlement can be reduced.

$$p = F/A \tag{5.2.10}$$

$$p = ks \tag{5.2.11}$$

where, F is the load (KN) transmitted from the total load of the upper covering soil, the tunnel structure of the underground expressway and the structure of the rail transit section to the bottom plate of the rail transit structure; A is the area of the bottom plate of the rail transit structure; P is the load strength (KPa); K is the coefficient of subgrade reaction; S is the settlement corresponding to P (mm).

From formulas (5.2.10) and (5.2.11), it can be seen that if A increases, then P decreases, when the coefficient of the subgrade reaction is constant, S will decrease.

$$s = \psi \frac{p_0}{E}(Z \cdot \bar{a}) = \psi \frac{\Delta p}{E} Z \tag{5.2.12}$$

where, $\psi$ is the empirical coefficient of the settlement; $p_0$ is the additional stress acting on the bottom plate of the structure in the rail crossing section; $\Delta p$ is the average additional stress above and below the top of the structure in the rail crossing section and the hard soil layer; $Z$ is the distance from the bottom plate of the structure in the rail crossing section to the top of the hard soil layer respectively; $\bar{a}$ is the average additional stress coefficient from the bottom of the structure to the layer of harder soil; $E$ is the compression modulus of the silt soft soil, which can be obtained by indoor testing.

It can be seen from formula (5.2.12) that, if $\Delta p$ and $p_0$ decrease, S will decrease.

(5) Based on the principle of pile-soil interaction and coordinated deformation, the applicant has found through research that it is only necessary to embed an inclinometer probe in the diaphragm wall of the retaining structure, and it is not necessary to embed another inclinometer probe beside the retaining structure of the foundation pit, which can reflect the horizontal deformation of the retaining structure of the foundation pit.

(6) The water in the covering soil directly infiltrates and erodes the roof and side wall of the tunnel structure of the underground expressway, so the roof and side wall of the tunnel structure as well as the bottom plate should be wrapped with waterproof coiled materials, and the roof should be coated with waterproof paint to prevent the corrosion of the steel bars in the concrete structure which would affect the mechanical properties of the structure; the upper water continues to infiltrate through the gap between the side wall and the retaining structure of the tunnel. The side wall and the bottom plate of the rail transit section structure are waterproof strengthening parts. We can learn from the rubber gasket commonly used in shield tunnel segments and the water swelling rubber strip, and use the rubber gasket and the water swelling rubber strip to waterproof the outside of the side wall and the bottom plate of the structure of the rail transit section.

**Problem 5.2.8**

With the construction of metro in large and medium-sized cities in China, many shield tunnels need to be built, but the segments of shield tunnels are prone to crack in the uneven transition section of stratum or structure. In order to learn how to prevent or deal with this kind of problem, it is necessary to study and improve the design and construction methods to prevent and control the segment cracking of the shield tunnel in the uneven transition section from the mechanical analysis of the interaction between the shield and the stratum.

**Solution** In view of the fact that the segments of shield tunnel in the existing stratum or transition section with uneven structures are prone to cracking, different grouting processes and types will produce different mechanical relations between shield and stratum. **The design and construction should be improved according to the coordinated control method of structural deformation.** On the basis of the design specifications, the layer of the effective bearing structure, control of the construction process and space stability should be formed in advance or in time or inherently. That is "space-time effect", especially the problem of space stability and control of the deformation coordination during the construction process. **To ensure that the force is transferred according to the design path, the unreasonable or even harmful transfer of the control force can avoid the collapse of the excavation surface.** According to the mechanical relationship between shield and stratum:

$$P + T = P_0 \tag{5.2.13}$$

where P is the mutual support force between the strata; T is the pressure borne by the shield segment; $P_0$ is the sum of the internal and external loads corresponding to the structural state.

It can be seen from formula (5.2.13) and Fig. 5.35a that $P_0$ is the sum of the internal and external loads corresponding to the structural state, which is approximately constant. To prevent the shield tunnel segment from cracking in the stratum or the transition section with uneven structures, a reasonable design and construction method should be adopted to increase the mutual supporting force P between the stratums and reduce the pressure T borne by the shield segment.

It can be seen from Fig. 5.35a that state ①: If the gap between the shield excavation layer interface and the installation segment is injected with hard slurry synchronously, the original layer has a tendency to move towards the gap but there is no moving space, the surrounding stratum will produce a supporting force P to prevent the movement of the original layer, which will inevitably reduce the pressure t of the original stratum on the segment; that is to say, the direction of the main stress of the stratum is shifted to the outside of the segment, which is favorable for the segment structure layer. State ②: If the gap is filled with inert slurry, the original layer and surrounding layer are not filled with the gap, resulting in a corresponding loosening and moving to the gap to further fill the gap, the surrounding layer will weaken the supporting force P of the movement of the original layer, and will jointly increase the pressure t of the stratum on the segment, that is, the direction of the main stress of the stratum tends towards the center of the segment structure, which is harmful to the layer of the segment structure.

Therefore, the design and methods of construction to prevent or treat the segment cracking of the shield tunnel in the uneven transition section of stratum or structure are as follows: (1) If the design and construction process of the shield tunnel reaches state ①, the purpose to prevent the segment cracking of the shield tunnel in the uneven transition section of stratum or structure can be achieved; (2) If the design and construction process of the shield tunnel is in state ②, it is necessary to fill the space again with double liquid slurry in time by means of Fig. 5.35b to control the deformation of the stratum, stabilize the interaction between the shield and the stratum, control or reduce the pressure t of the original stratum on the segment, and repair the segment cracks again, in order to deal with the cracking phenomenon of the shield tunnel segments in the non-uniform transition section of the stratum or structure.

**Problem 5.2.9**

Due to the planning requirements, some mountain highway tunnels built in China need to cross high, steep critical slopes. The code for designing them emphasizes the design and construction according to the New Austrian Tunneling Method (NATM),

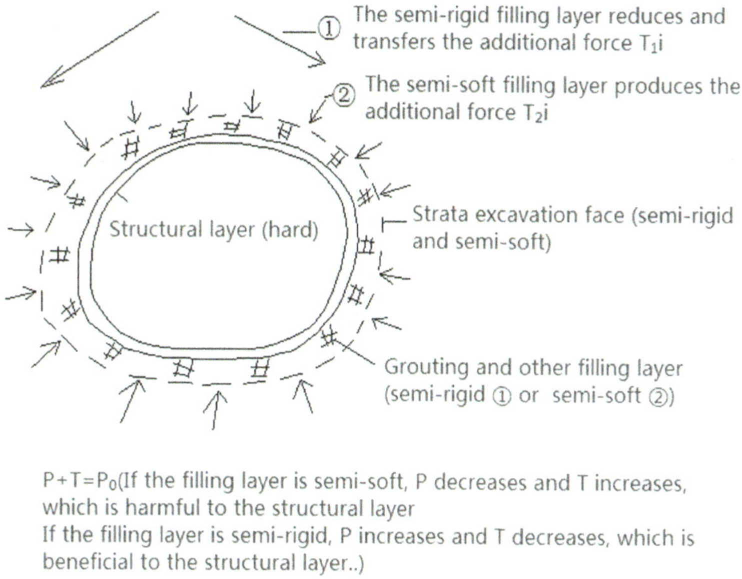


(a)

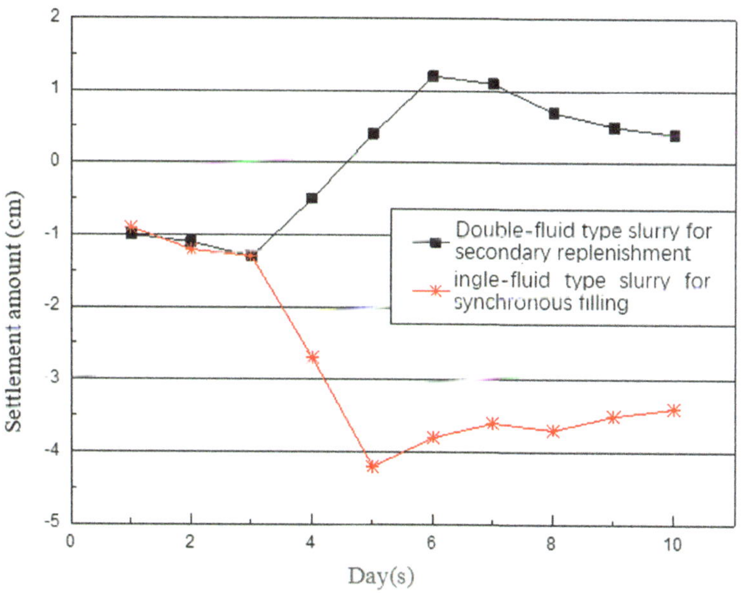


(b)

**Fig. 5.35** **a** Diagram of the interaction between shield and stratum in the uneven transition sectionXQ; **b** Comparison of the deformation of the formation controlled by the single-ring single-liquid type and double-liquid type slurry filling

**Fig. 5.36** Photo of the collapse of highway tunnel portal construction in a mountainous area

but the implicit application condition of the NATM is the overall stability of the mountain, only the stability of the surrounding rock is studied. When the mountain highway tunnel needs to cross a high, steep critical slope of the portal due to the planning requirements, some design and construction personnel often neglect the stability of the high, steep slope of the portal, sometimes causing the collapse of the portal construction or even casualties, as shown in Fig. 5.36.

**Solution** In view of the problem of stable equilibrium between the design and construction mechanics of the high, steep critical slope at the entrance of some mountain highway tunnels, when the rock mass of the high, steep critical slope at the entrance of the tunnel is poor, collapse of the tunnel construction often occurs; at this time, the method of the control of the pre-equilibrium and stability must be adopted to solve this kind of problem of collapse during the construction process. That is, (1) The pre-equilibrium and stability control measures of retaining the back pressure at the foot of the slope are adopted to stabilize the critical slope of the high, steep tunnel portal as a whole, as shown in Fig. 5.37a; (2) The scheme for the control of the pre-equilibrium and stability of the leading pipe shed + arch support zero excavation into the tunnel is adopted to further stabilize the critical slope of the tunnel portal as a whole, as shown in Fig. 5.37b; (3) Adopt the approximate full cross-section construction method of short step reserved core soil to shorten the longitudinal length of the tunnel excavation and support construction, place the tunnel excavation and support construction close to the plane mechanical problem required by the design, so as to reduce the disturbance effect on the high, steep critical slope of the portal, and thus let it play the role of an auxiliary stabilizing the high, steep critical slope of the portal, as shown in Fig. 5.37c. When the face of the tunnel face is stable in the process of

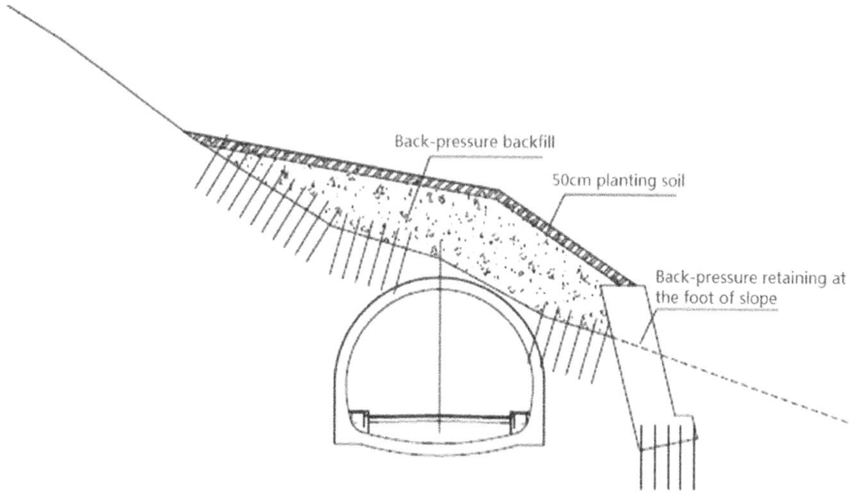


1) Elevation diagram

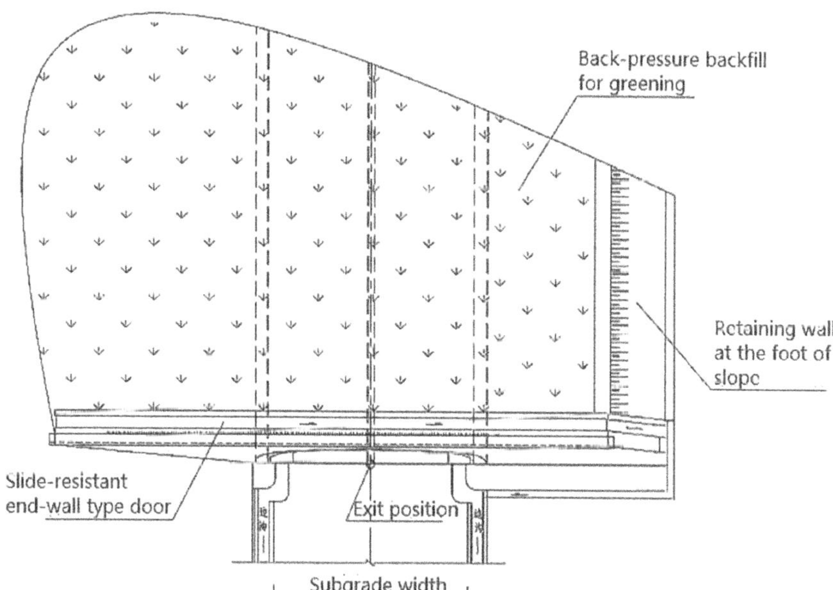


2) Plan diagram

(a)

**Fig. 5.37** **a** Control measures for pre-equilibrium and stability for retaining the back pressure of the high, steep critical slope at the tunnel portal; **b** Control scheme for pre-equilibrium and stability of zero excavation into the tunnel concerning the advance pipe shed + arch support for the high, steep critical slope at the tunnel portal; **c** Diagram of the approximate full cross-section construction method of short step reserved soil core

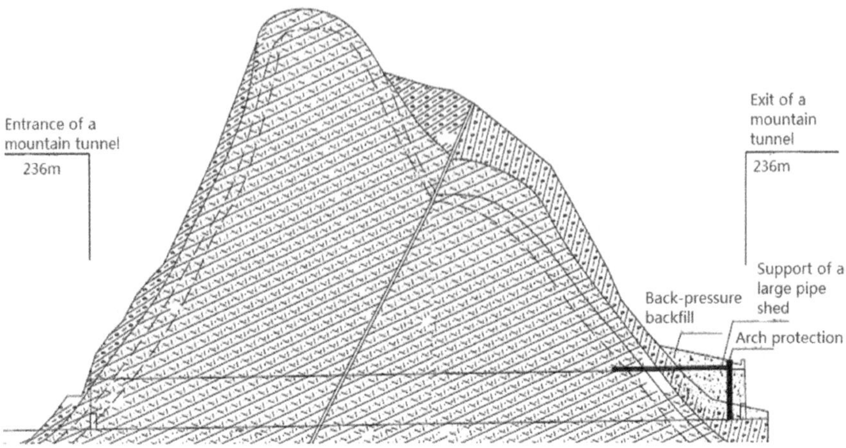


(b)

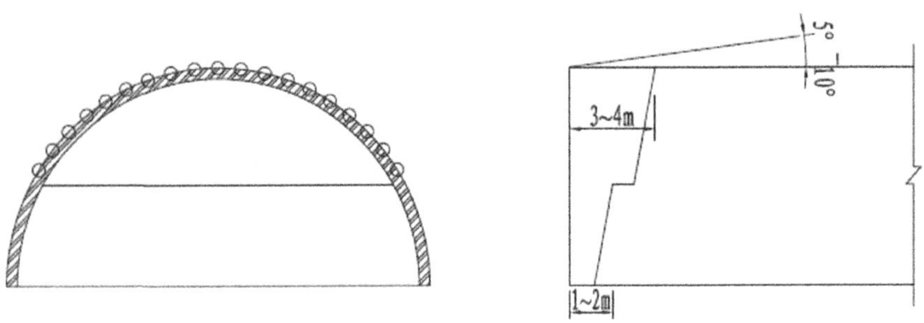


(c)

**Fig. 5.37** (continued)

tunnel construction, it is advisable to adopt the construction method of short step reserved core soil approximate full cross-section, as shown in Fig. 5.37c. Tunnels with class III, IV and V surrounding rock adopt short steps and reserve core soil for the upper and lower steps under the strong pre-support. The upper and lower steps are parallel to each other by 1–2 m. They should follow the strong initial support in time and bear all loads to ensure the stability and equilibrium of the tunnel structure construction process.

**Problem 5.2.10**

At present, there are two main types of tunnel construction methods in China: one is blasting construction, which is applied in various kinds of tunnel engineering. There are problems, such as uncontrolled size differences, construction risks and excessive blasting vibration. The blasting vibration is limited in the construction process of the new medium and short length rock tunnel near high-speed railways, highways, crossing or approaching, and buried shallow at the bottom of an area crossing water, so the blasting construction is not suitable. The other is mechanized construction, which can be divided into shield construction and cantilever tunneling machine construction. The former adopts a shield machine with a high construction cost and inconvenient transportation, which is suitable for large-scale projects such as long-distance tunnels; the latter adopts a cantilever-type road-header for construction, when encountering hard rock tunnels, the pick wear is accelerated and the efficiency of the construction is reduced. For the non-explosive construction of a hard rock tunnel, there are other methods, such as the method using mainly a full cross-section tunnel boring machines (TBM), the method using a cantilever tunnel boring machine, the method using a milling and excavating machine, the splitting method, the method using a hydraulic impact hammer, the static crushing method and so on. The TBM is not suitable for medium and short length tunnels because of its high cost. The mining efficiency of the other methods depends on the hardness of the tunnel rock. When the hardness is too great, the efficiency will be too low or hard rock will not be excavated. It is difficult to meet the economic efficiency and safety risks of the construction of high-speed railways, near highways either crossing or approaching, as well as the newly built medium and short length rock tunnel buried shallow at the bottom of the area crossing water.

**Solution** Aiming at the problem of economy and efficiency of medium and short length hard rock tunnel excavation, there is a new mechanical excavation method, which can give consideration to both economy and efficiency at the same time. The specific steps are as follows: First, several parallel vertical grooves are cut on the tunnel face every certain distance, and each vertical groove makes the rocks on both sides form a certain depth of a free face; then, the crushing equipment is used to squeeze the free face on both sides of the vertical groove, so that the rock mass on both sides of the vertical groove is broken, and the tunnel excavation is realized (Fig. 5.38).

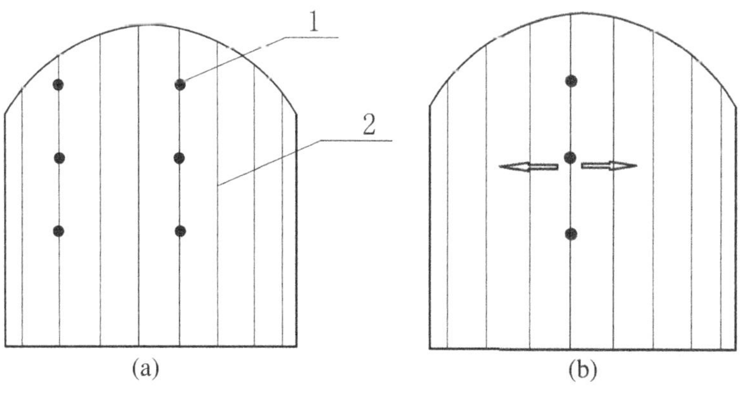


**Fig. 5.38** **a** Layout of vertical groove and fracturing point; **b** Diagram of the construction sequence of vertical groove crushing

The principle is as follows: There are many micro cracks inside the complete hard rock tunnel. In the case of the complete hard rock closure, the energy accumulation is very great. The micro cracks are not conditionally released. Hydraulic fracturing is very difficult, and the economic efficiency of tunnel excavation and support is very low, which cannot meet the production needs. On the basis of complete hard rock, many parallel vertical grooves are cut every 50–60 cm. The free surface at both ends of the vertical groove creates displacement boundary conditions for the release of micro cracks inside the hard rock, and the integrity of the energy accumulation is reduced to local, which further creates good conditions for a hydraulic fracture, accelerates the progress of production and controls the safety risks. A certain point on the vertical groove of the crushing equipment is used as the fracturing point to apply pressure. At that time, the rock mass at the back of the free face is transformed into the shear force on the rock mass, and the shear strength of the rock is much smaller than the compressive strength. When the extrusion pressure reaches a certain degree, the micro cracks inside the hard rock expand, and the existence of the free face makes it possible to displace, so the rock mass breaks. Constantly crush different points on the face, complete the excavation of the whole face, and then cycle the excavation.

**Problem 5.2.11**
At present, China's road, rail and other comprehensive transportation ways are developing rapidly, and the phenomenon of three-dimensional intersections of highway and railway tunnels in mountainous areas is increasing. When the mountain geology is general and close to the intersection, how can the design and construction of multiple cross-mountain highway and railway tunnels be dealt with? In order to avoid the construction risk of the new highway tunnel and improve the stress of the operating highway tunnel, the design and construction methods of multiple cross-mountain highway and railway tunnels are developed around this goal.

**Solution** In view of the existing method of the three-dimensional intersection treatment of highway and railway tunnels in mountainous areas, it is easy to ignore the analysis of the structural mechanics of the interaction between the stratum and the tunnel, and there is no macroscopic qualitative control method.

The first step is to pre-strengthen the secondary lining of the existing highway tunnel, as shown in Fig. 5.39b.

In a range of 60–70 m in the longitudinal direction of the secondary lining of the existing highway tunnel at the lower part of the three-dimensional intersection of the tunnel, the MPC glue or epoxy concrete is used to paste 20 cm wide and 6–8 mm thick steel plates onto the secondary lining of the existing highway tunnel every 50–100 cm, and the secondary lining of the existing highway tunnel is pre-reinforced.

The second step is to conduct the design and construction of controlled blasting for the excavation of a new highway tunnel, as shown in Fig. 5.39a.

According to the *Safety Code for Blasting* (GB6722-2011), the maximum allowable vibration speed of the traffic tunnel is less than 10 cm/s, and the recommended vibration speed of a highway tunnel for many years has been less than 5 cm/s. The

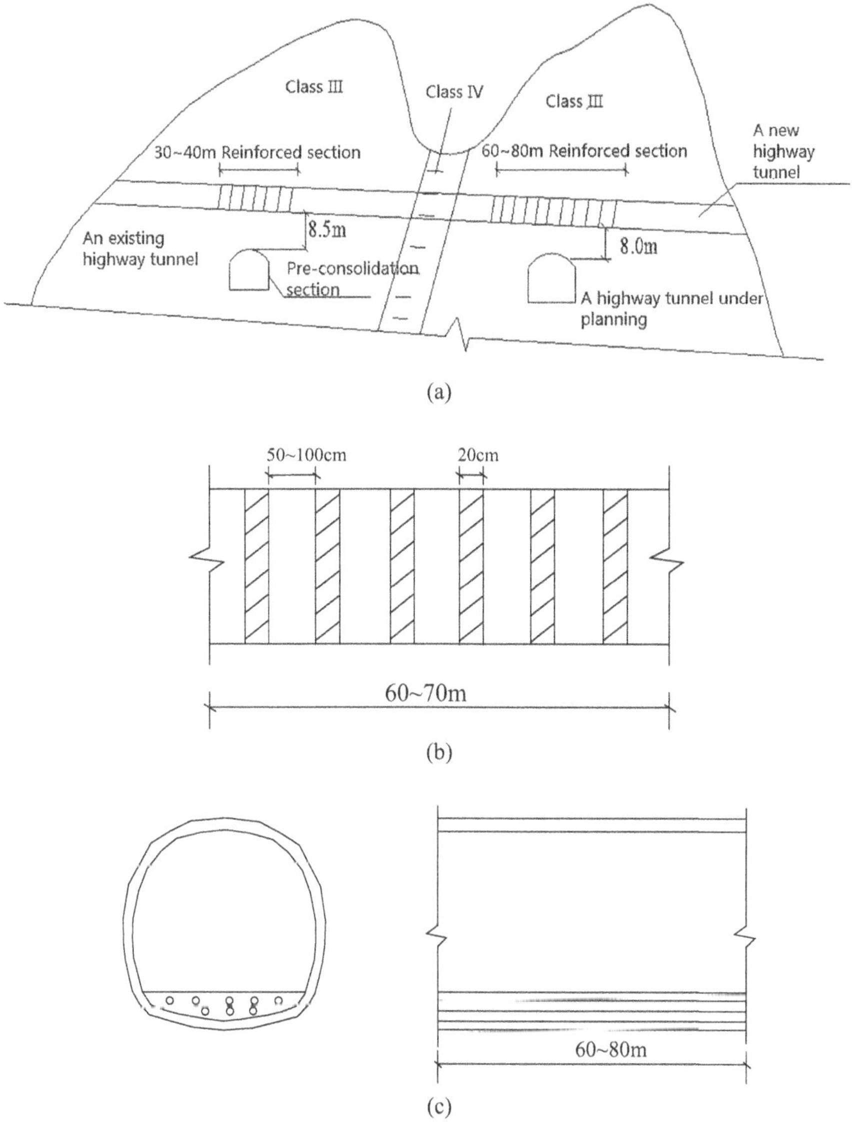


**Fig. 5.39** **a** Cross-section diagram of three cross highway tunnels; **b** Cross-section diagram of the pre-reinforcement of the secondary lining of the existing highway tunnel; **c** Diagram of the composite structure of the inverted arch longitudinal beam of reinforced concrete for the secondary lining of the reinforced section of a new highway tunnel

results of the comprehensive calculation show that: Within 15 m (i.e. 60–70 m) before and after the reinforcement section of the new highway tunnel in the upper part of the existing highway tunnel, and within 60–70 m of the planned highway tunnel crossing, the full cross-section excavation with a footage of 1.0 m or the step excavation with a footage of 1.5 m is carried out, and the single section dosage is controlled at 5 kg to meet the above requirements.

In the third step, during the excavation of the new highway tunnel, the initial support should be done in time, and the construction of the strengthening section should be done well, as shown in Fig. 5.39a, c.

For each 1.0–2.0 m excavation of the upper newly-built highway tunnel, the initial support should be done in time to ensure the stability of the surrounding rock of the upper newly-built highway tunnel, and the secondary lining of the newly-built highway tunnel should be done under the condition of following the initial support as required.

In a range of 30–40 m above the existing highway tunnel and 60–80 m above the planned highway tunnel, it is beneficial to improve the stress of the existing highway tunnel, avoid the construction risk of the planned highway tunnel and improve the stress of the planned highway tunnel to improve the inverted arch composite structure of the secondary lining reinforced concrete of the reinforced section of the new highway tunnel.

## 5.3 Method for the Treatment of the "Bump at Bridgeheads" for the Soft Soil Foundation of Highways

The soft soil structure belongs to the problem of ash boxes. There are not only clear parts (such as force) that can be calculated theoretically, but also most unclear parts (such as deformation or settlement) that have large errors of theoretical calculations, so it is more effective to adopt control measures in actual projects. This is also the reason why the author compares the difference between Jiangsu lime soil subgrade of soft soil and the Zhejiang soft soil slag subgrade, and makes statistics regarding the results of about 50 research projects on Zhejiang soft soil slag subgrade carried out in the past 20 years, which is not ideal for the effect of controlling the "bump at the bridgehead" of highway soft soil slag subgrade. It is also the reason why the author has successfully solved the problem of controlling the "bump at the bridgehead" of highway soft soil subgrade by combining theoretical calculations with structural control measures in 9 highways, such as those in Jiaxing and Ningbo.

There are many methods to deal with and analyze the soft soil subgrade of highways, sometimes only to improve them. Why hasn't the problem of the "bump at the bridgehead" been solved? In the world, it is generally believed that the calculation of the stress of the soft soil foundation is more accurate and the error in the calculation of the deformation is larger, which is in contradiction with the relationship of the stress deformation function y = f(x). The key problem is that the state of deformation of the soft soil in the process of the mechanical analysis of the soft soil subgrade is different from that of the structure in the process of engineering mechanics.

In view of the problems existing in the existing theory of analysis and the method of the treatment of highway soft soil subgrade, the author studies the mechanical properties of soft soil subgrade, especially the rheological properties of soft soil subgrade, adheres to the principle that the theory of mechanical analysis and the method of practical application of highway soft soil subgrade match the requirements of engineering mechanics, and he proposes for the first time the methods such as using the lower threshold control of soft soil rheology and using the lower partition under the subgrade. On the basis of the current code of design, this paper puts forward the method of designing and the complete set of technologies to control the stability of the state of deformation due to stress and the transition of the longitudinal settlement of the soft soil subgrade of the highways, and solves the problem of the "bump at the bridgehead" that has puzzled the traffic field for decades. This is not so much a science as a technology as a technology as a science. The most important purpose is to reasonably complete the design and construction of geotechnical engineering. Mathematics and mechanics are just the means.

The author reveals the mechanism that affects the "bump at the bridgehead" of the highway soft soil foundation:

I. The consolidation, secondary consolidation and rheological properties of soft soil foundations are related to the level of stress of the foundation. According to the subgrade coring test (Figs. 5.40, 5.41, 5.42), the soft soil increases rapidly

**Fig. 5.40** Rheological test of highway soft soil

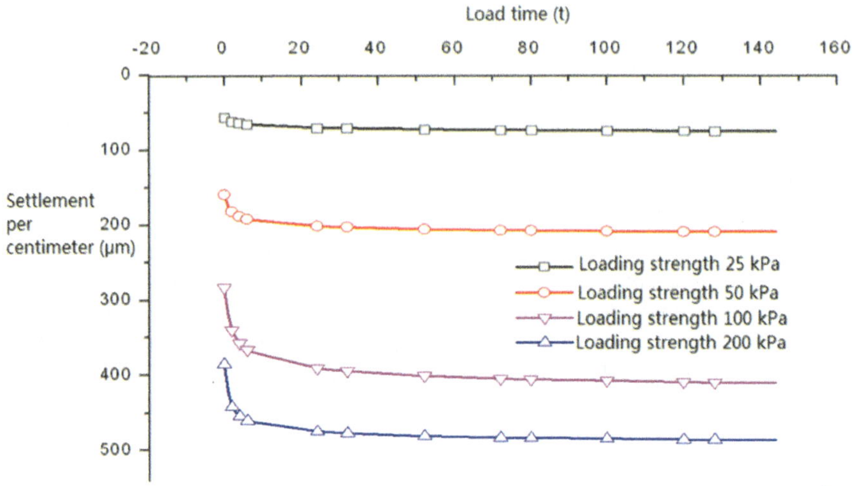


**Fig. 5.41** Rheological curve of highway soft soil

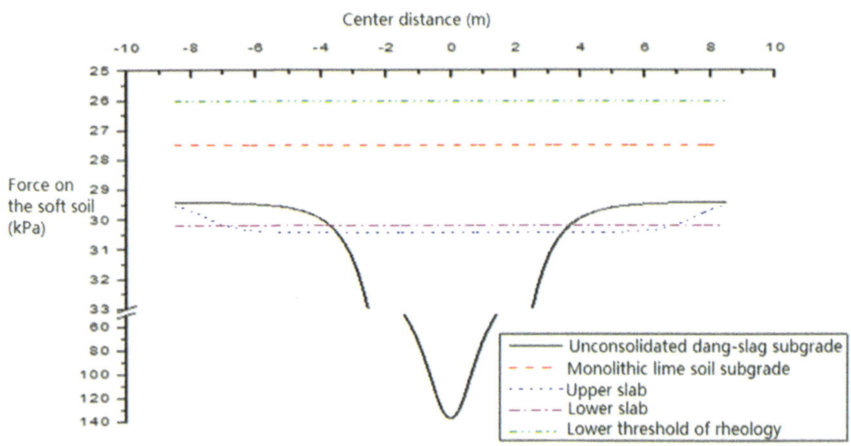


**Fig. 5.42** Diagram of the stress of the interface between the highway soft soil subgrade and the foundation

with time at the initial stage of settlement, then the increasing trend gradually decreases and tends to become stable. When the stress level is 25, 50, 100 and 200 kPa, the influence of rheology on the settlement of the soft soil can be ignored; when the stress is 50, 100, 200 kPa, the rheology has a great influence on the settlement of the soft soil. The core test has a lateral limit and only vertical deformation. The actual settlement of the bridgehead subgrade has a lateral compression deformation without a lateral limit. Then, when the stress of the soft ground reaches a certain value, the settlement of the bridgehead subgrade will not be stable. Through systematic research and testing, it has been found that there is a phenomenon of a "lower threshold of soft soil rheology". When the stress level of the soft soil foundation (caused by the weight of the soil foundation itself plus the live load of vehicles, including the vibration and impact of heavy vehicles) $\geq$ the lower limit of soil (soft clay) rheology, secondary consolidation rheology of the soil will occur. The settlement of the soft soil subgrade will be very large after construction, and the influence of soft soil rheology on the subgrade settlement cannot be ignored.

According to the *Technical Standard for Highway Engineering*, the typical parameters of automobile tests are selected, with the vehicle weight of 550 kN (about 55t), four middle and rear wheels, the landing width and length of 0.6 m $\times$ 0.2 m, two front wheels, and the landing width and length are 0.3 m $\times$ 0.2 m, the overall dimension of the vehicle is 15 m $\times$ 2.5 m. The density of the slag subgrade material is 2,000 kg/m$^3$, the density of the lime soil subgrade is 1,730 kg/m$^3$, and the lower limit of the rheological property of the soft soil subgrade is 26 MPa. When the subgrade filling height is 1.5 m, the analysis of the stress of the interface between the soft soil subgrade and the foundation is shown in Fig. 5.42. Among them, the proportion of lime soil subgrade is small, the subgrade has stability, the stress at the base of the soft soil road is less than the "lower threshold value of the soft soil rheology", which avoids the adverse effect of a concentrated dynamic load of vehicles, and there is no phenomenon of a "bump at the bridgehead"; while if the ratio of slag road to subgrade is large, the subgrade has no stability, the stress at the base of the soft soil road is greater than the "lower threshold value of soft soil rheology", and the impact of a concentrated dynamic load of vehicles is large, which results in a "bump at the bridgehead".

II. The soft soil of the highway soft soil subgrade has the characteristics of a large void ratio, high compressibility, low strength and high sensitivity, which can easily cause post-construction settlement of the subgrade. The core problem is that the model of the design and analysis of the highway soft soil subgrade is inconsistent with the actual state of the stress and deformation of the soft soil slag subgrade (Fig. 5.43). The ideal subgrade belongs to the continuum model, while the dang-slag subgrade belongs to the discrete subgrade. The main difference between the continuum and the discrete is that: The discrete can bear the pressure, but basically cannot bear the tension and the moment; the continuum can bear the pressure, tension and moment. In the actual project, the vehicle load is a typical point dynamic load, which will form an obvious tension bending area at the bottom of the subgrade, not an ideal pure pressure area. At this time, if the uniform load is treated according to the current specifications, and the continuum model is still used, it will produce a lot of errors.

The main difference between the Jiaxing highway and Suzhou highway is the stability of the subgrade. The No.1 highway adopts the dang-slag subgrade, which belongs to the loose subgrade. It cannot effectively play the role of hard shell of soft soil foundation (including artificial) and is not conducive to the stress of the soft soil subgrade. The settlement and deformation of the bottom

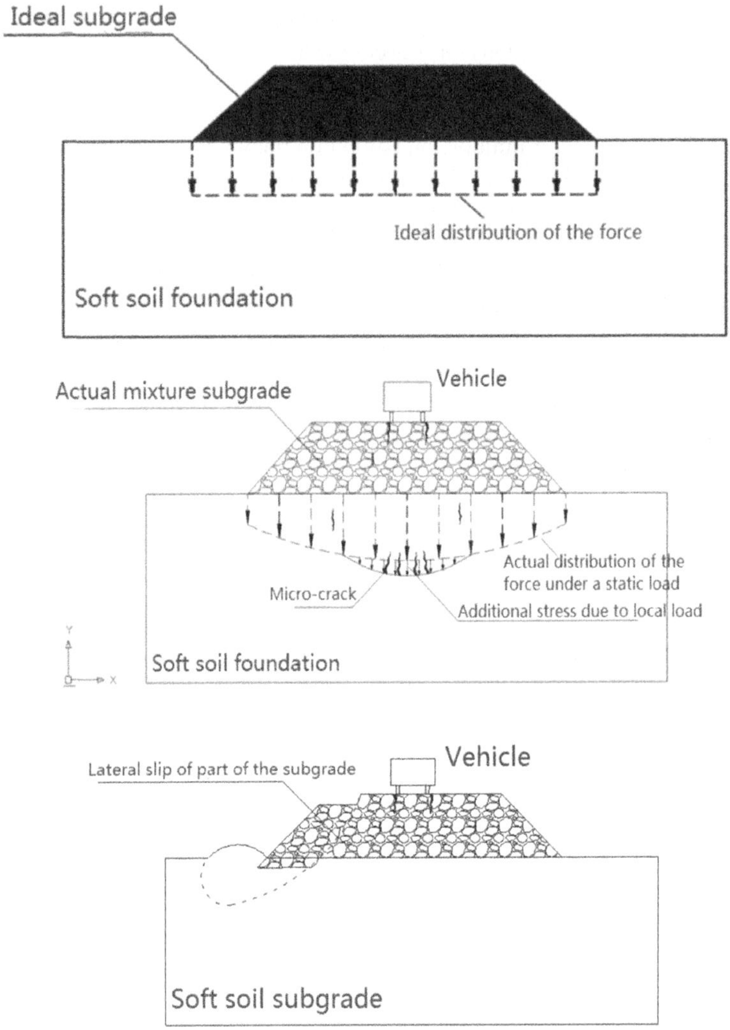


**Fig. 5.43** Design model of highway soft soil subgrade and the actual state of the soft soil slag subgrade

of the subgrade is of the convex belly type, which does not meet the model requirements of the design specifications, the accumulated settlement of the subgrade at the bridgehead is about 1.0–1.5 m, and there is the phenomenon of a "bump at the bridgehead". On the contrary, the No.2 highway adopts the lime soil subgrade, which is a stable subgrade. It can not only effectively play the role of the hard shell layer of the soft soil foundation (including artificial), but it can also uniformly distribute the concentrated load of the pavement, and basically control the stability of the state of deformation due to the stress of the soft soil subgrade. The settlement deformation at the bottom of the subgrade is the flat type, which meets the requirements of the model of the design specifications. The accumulated settlement of the subgrade at the bridgehead is less than 5 cm, and there is no "bump at the bridgehead" after regular maintenance or overlay every year; it has been revealed that insufficient subgrade stability and the stress at the bottom of the soft soil road that is greater than the "lower threshold of soft soil rheology" are the main reasons for the phenomenon of a "bump at the bridgehead" of the highway soft soil foundation, and also the reasons for the uncontrollable long-term compression settlement after the construction of the dang-slag subgrade.

The effect of the layer of the bearing structure formed by soft soil subgrade without a lower partition and improved soil is generally unstable. At this time, the state of the stress and deformation of the soft soil subgrade sometimes changes or even loses stability, especially the problem of a "bump at the bridgehead", and sometimes only improves. For example, if the highway composite foundation basically covers the whole subgrade, it is close to the problem of a plane strain, but the cost is very high. If the coverage rate is small, the effect of controlling the subgrade settlement is poor. The soft soil subgrade of improved soil or slag combined with lower partition and pile transition technology can give full play to the role of the soft soil foundation (including artificial) hard shell layer to form an effective layer of the bearing structure, and can basically control the post-construction settlement of the soft soil subgrade, so as to solve the post-construction settlement of the soft soil subgrade, especially the problem of a "bump at the bridgehead" (Figs. 5.44 and 5.45).

The author first put forward the threshold of the lower limit of the soft soil and the first use of the lower partition to control the settlement of the soft soil subgrade. According to the coordinated control method of structural deformation, the method of designing the soft soil subgrade has improved. ① The subgrade can adopt EPS, foam concrete, improved soil subgrade, adding the lower baffle or frame to the dang-slag subgrade to control the stability of the

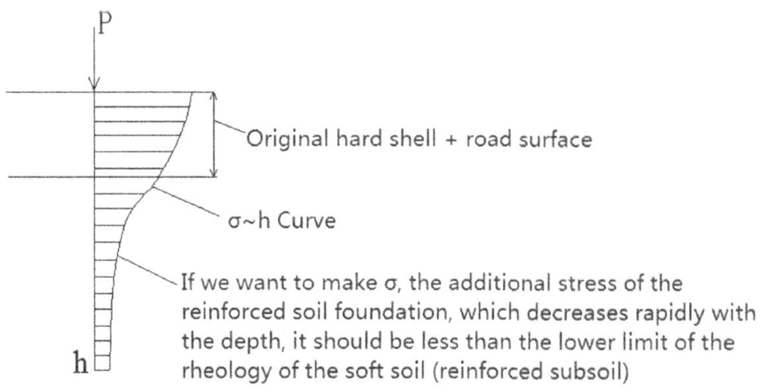


**Fig. 5.44** Relationship between the soft soil structure and the distribution of stress

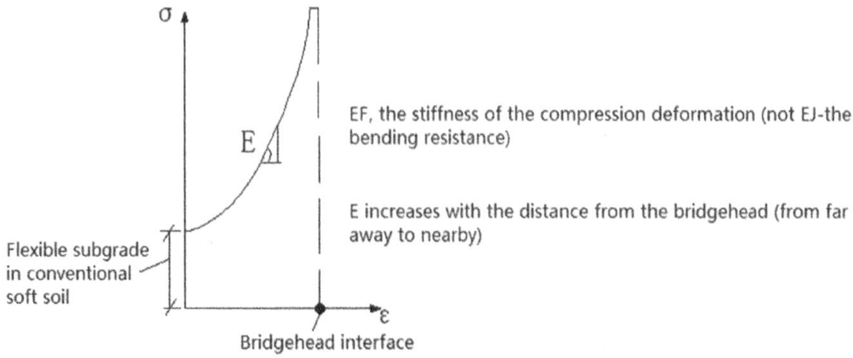


**Fig. 5.45** Variation relationship of the stiffness of the compression deformation of the bridgehead foundation

soft soil subgrade under the condition of stress and deformation. In addition, the change of the extension slope of the subgrade (bottom plate) from the bridge end should be controlled within 0.4% or the inclination should be controlled within 0.2–0.3°; ② When the stress value of the soft soil subgrade is smaller than the lower threshold of the soft soil rheology, the post-construction settlement of the soft soil subgrade can be controlled; when the stress value of the soft soil subgrade is greater than the lower threshold of the soft soil rheology, engineering measures should be taken, such as adding a lower partition board or adding a lower partition board to the light subgrade or slag subgrade Only with the combination of frame and pile can the settlement of the soft soil subgrade be controlled within the allowable range. Therefore, in order to solve the problem of post-construction settlement of the highway soft soil subgrade, according to the characteristics of the local soft soil foundation, the stress should be controlled within the lower threshold of the soft soil rheology, or corresponding engineering measures should be taken to control it.

The author has developed a complete set of technical systems for the combination of a lower partition or settlement transition pile to treat the "bump at the bridgehead" of the highway soft soil foundation:

I. Technology for the treatment of the combination of light material and lower partition of the soft soil subgrade
[A case of a "bump at the bridgehead" treatment of a highway project in Taizhou] This is a project for widening the old bridge. The height of abutment is 2.4 m and the thickness of the soft soil is 39.5 m. In 1999, the old bridge was opened to traffic. The new bridge was completed in 2009. The lightweight material was directly filled on the lower partition board, and the light foam bead concrete was made up of foam beads (EPS particles), medium coarse sand, gravel, cement and water. The design of the composition was made by construction and mixing. The materials used in this test project were selected nearby. Foam beads (EPS particles) are spherical particles of polystyrene, and admixtures were used as

early strength superplasticizers, micro silica fume and flyash. Bubble concrete is a kind of light-weight material, which is formed by mixing the curing agent (cement), water, bubbles and other admixtures in a certain proportion. It has many characteristics, such as light weight, heavy weight and adjustable strength, high fluidity, stability after curing, good construction performance, durability and superior environmental protection. The polyethylene bubble block (called EPS block) of the finished block was replaced with the soil behind the abutment to reduce the pressure of the earth behind the abutment (as shown in Fig. 5.46). After 4 years of operations, the total settlement of the new bridge subgrade was 4.75 cm; the old bridge adopted the traditional slab technology approach, and there was still settlement of the bridgehead subgrade. The total settlement of the subgrade in 4 years was 11.23 cm, which exceeds the settlement of the widened bridgehead soft foundation.

II. Dang-slag subgrade and lower partition + pile foundation transition treatment technology

[A case of a "bump at the bridgehead" treatment of a highway project in Zhenhai] The land-form of a highway project in Zhenhai District is marine plain, and the whole line is a section of soft soil. The layout of the subgrade cross-section of the project is: 4 m sidewalk +3.5 m non motor vehicle lane +12 m motor vehicle lane +5 m central separation belt +12 m motor vehicle lane +3.5 m non motor vehicle lane +4 m sidewalk = 44 m. The soft foundation treatment is mainly the bridgehead section. The subgrade filling of the bridgehead section is relatively high, 2–3.5 m. In order to avoid the phenomenon of a "bump at the bridgehead" and ensure driving comfort, a pre-stressed pipe pile combined with a reinforced concrete frame was used for the soft foundation treatment. See Fig. 5.47 for the profile diagram and plane layout of the lower partition treatment of the highway soft soil subgrade.

The treatment of a soft foundation at the bridgehead can be divided into two cases: a reinforcement section and a transition section. ① The reinforcement section: Reinforce the bridgehead within 10.4 m, and the post-construction settlement of the treated road surface within the design service life should not be greater than 10 cm, so as to avoid a "bump at the bridgehead". The soft foundation is deep-treated with pre-stressed pipe piles, which are arranged in a square with a spacing of 2.6 m. A 1.2 m-wide vertical and horizontal reinforced concrete frame beam is set at the top of the pipe pile, which not only ensures the integrity of the subgrade stress and deformation, but it also makes full use of the overall stress of the soil between the piles and the pipe pile. According to the soft foundation of each bridgehead, the length the pile of the reinforcement section is determined by a theoretical calculation. ② The transition section: In general sections of the project, a soft foundation treatment should be carried out by reducing the filling height of the subgrade, using a surcharge preloading combined with a pre-throwing height to control the post-construction settlement of general sections within the design service life of the pavement of not more than 30 cm. The transition section is set between the reinforcement section and the general section, and the transition rate

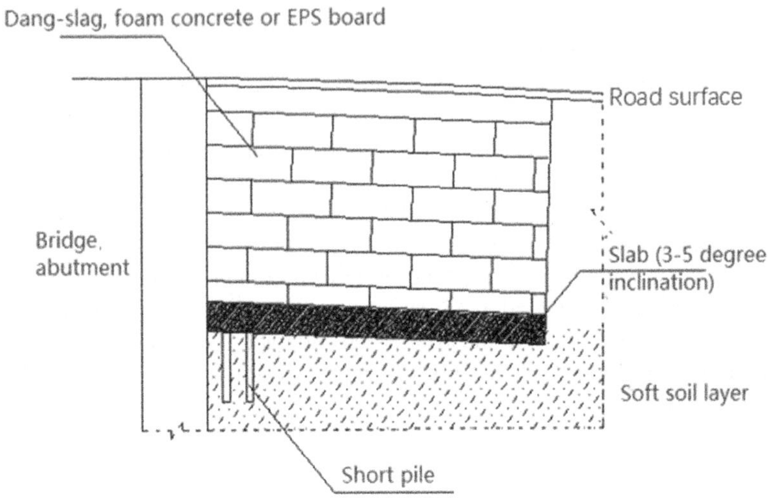


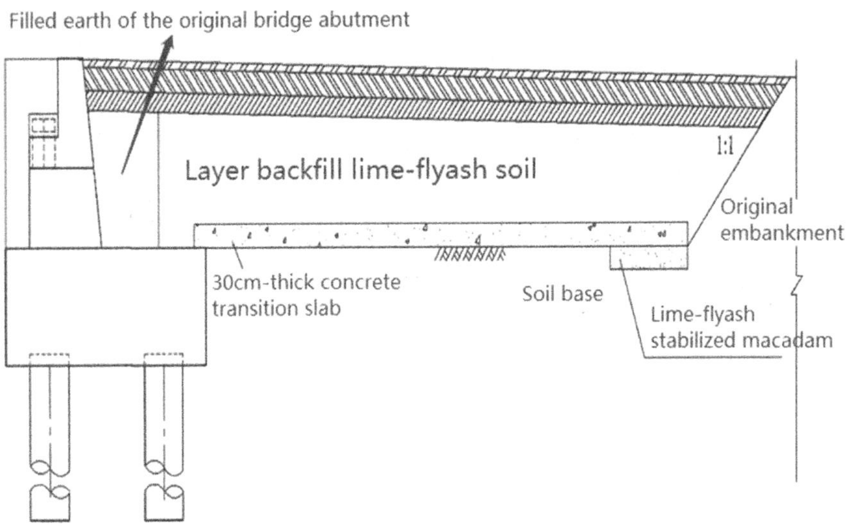


**Fig. 5.46** Profile of the light material and lower partition treatment for the soft soil subgrade of a highway

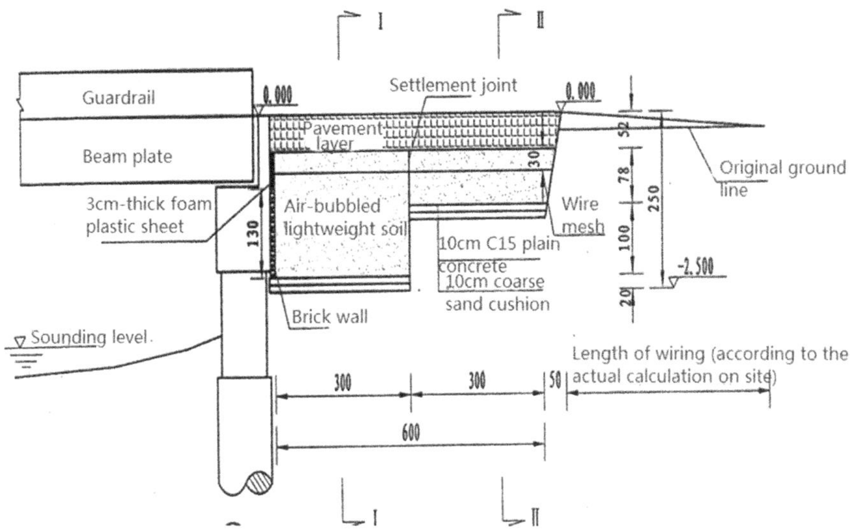


**Fig. 5.46** (continued)

of the differential settlement should be $\leq 0.5\%$. In the section of the soft foundation treatment with pre-stressed pipe piles, the transition section is generally set in three ways: a variable pile length, a variable pile distance, or a variable pile length and pile distance at the same time. In order to ensure the integrity of the stress and deformation of the subgrade, a reinforced concrete frame beam is used as the lower partition of the subgrade in the design of this project. The farther away from the bridgehead, the lower the filling height of the subgrade. In order to ensure the soil arching effect of the subgrade between the frame beams, it is not suitable to increase the spacing of the piles. Therefore, in this design, the spacing of the piles and the size of the frame beam are kept unchanged, and the transition is realized by changing the length of the piles. Four transition sections are set in this design. According to the soft foundation condition of each bridgehead, the length of the piles of the first, second and third transition sections are determined by means of a theoretical calculation. The fourth transition section is not provided with pre-stressed pipe piles.

III. Technology for the treatment of the composite subgrade such as lime soil or subgrade grouting

[A case of a "bump at the bridgehead" treatment of the Ouhai section of national highway 104] The whole process of the Ouhai section of national highway 104 in Wenzhou is 10 km, which belongs to a typical soft soil geological section. After its reconstruction and opening to traffic in 2000, due to the settlement of the soft foundation after construction, the phenomenon of a "bump at the bridgehead" was very serious. There are 13 bridges in the whole line, and the bridgeheads have to be paved every year on average. In order to effectively

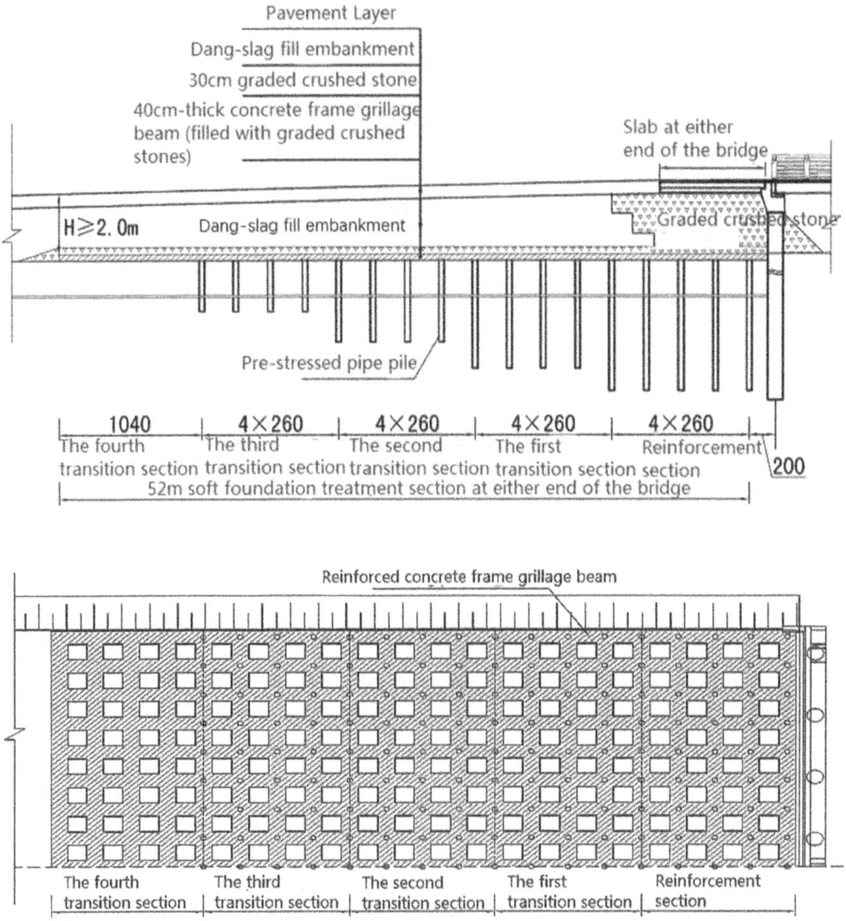


**Fig. 5.47** Profile diagram and plane layout of the lower partition treatment of the highway soft soil subgrade

solve the problem of "bumps at bridgeheads", the second bridge was taken as the experimental project in 2008, and four different methods were adopted to treat the four bridgeheads. ① The "DGR I method" Technology for the quick repair of the Deep Grouting Reinforcement abutment backfilling (Fig. 5.48). The length of the grouting treatment was 50 m, the grouting hole diameter was 5 cm, the grouting hole was arranged in five rows horizontally, the spacing of the longitudinal holes was 2–4 m, the grouting depth was 10 m, (the depth was determined according to the requirements for the foundation bearing capacity and the degree of the softness of the soil), the grouting pressure was 0.5–1 MPa, and the traffic was opened 12 h after the grouting. ② Remove the bridgehead pedal and conduct joint treatment by grouting on the filling surface of the abutment back. Construction process: the same as above. ③ Grouting at the bottom

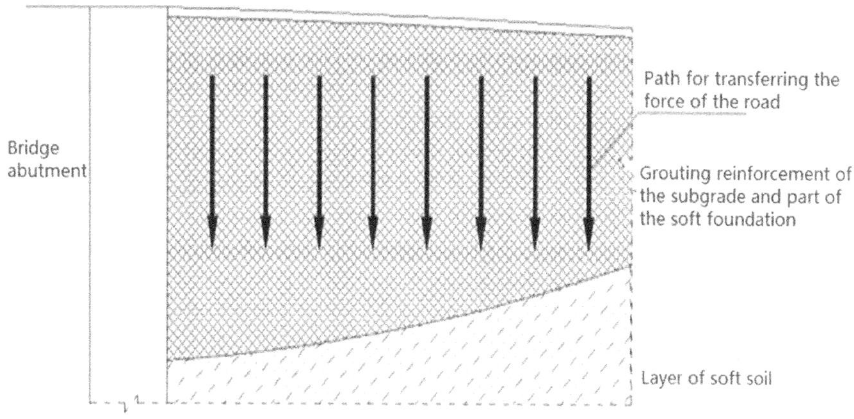


**Fig. 5.48** The state of the stress and deformation of the dang-slag subgrade after adding ash or grouting

of the approach slab and vertical reinforcement at the end of the baffle. Grouting reinforcement should be carried out from the backfilling of the abutment. The grouting range was the length of the approach slab plus 1.5 m, the diameter of the grouting hole was 5 cm, the grouting hole was arranged in a triangular shape, the spacing was 150 cm × 150 cm, the shallow layer grouting at the bottom of the approach slab, the vertical grouting depth at the end of the approach slab was 10 m, the grouting pressure was 0.5–1 MPa, and the traffic was opened 12 h after the grouting. ④ Cement stabilized layer was used for the connection with the slope, and asphalt concrete was used for surface layer. This method is mainly used as a comparative scheme. For the observation of the settlement of the connection section of the trial bridgehead slope, three horizontal observation points were respectively arranged at 1, 5 and 9 m away from the abutment. The observation after one year's use showed the total drop to be about 1 cm, which shows that the effect of the method of deep grouting reinforcement was obvious.

Since the application in 2009, all the projects completed by the author have been successful cases, breaking the view that the treatment of "bumps at bridgeheads" of highways with a soft soil foundation is a worldwide problem in the history of Zhejiang Province, laying a good foundation for summarizing and popularizing the new system of a method for the designing for the treatment of "bumps at bridgeheads" of highways with a soft soil foundation, and has achieved good social, economic and ecological benefits; moreover, it has a wide range of application value.

**Problem 5.3.1**
The problem of post-construction settlement of the existing highway soft soil subgrade is relatively prominent, especially for the embankment filled with a mixture of soil and rocks, which is reflected in the longitudinal and transverse cracks in the pavement; however, for the integral subgrade such as the lime soil subgrade and the subgrade of a mixture of soil and rocks in the later stage of grouting, the phenomenon of longitudinal and transverse cracks in the pavement is rare; the core problem of the post-construction settlement of the highway soft soil subgrade is to control the integrity of the state of stress and deformation of the subgrade. How to solve the problem of an integral control of the state of the stress and deformation of the new road subgrade in places with more coastal mountains and less land and abundant slag economically can avoid the repeated problems of longitudinal and transverse cracks in the surface of the roads, save construction and maintenance funds, and achieve more in one stroke!

**Solution** In view of the problem of the integrity of the state of the stress and deformation of the existing new slag subgrade filled with a mixture of soil and rocks, the frame and other measures were adopted between the bottom layer of the subgrade and the top layer of the soft soil foundation to control the integrity of the state of the stress and deformation of the newly built slag subgrade filled with a mixture of soil and rocks, so as to avoid the longitudinal and transverse cracks of the pavement, as shown in Fig. 5.49a–c.

**Problem 5.3.2**
There are many post-construction settlement phenomena of soft soil subgrade, especially the relatively poor characteristics of soft soil subgrade near mountains and water, and the problem of a lateral slip of the soft soil subgrade is more prominent. However, there are many methods to treat and analyze the foundation for the existing soft soil subgrade. Why are the successful cases not widely applicable? The two reasons that can easily be ignored are: (1) The main difference between the subgrade continuum and the dispersion: the dispersion can bear the pressure, but basically it cannot bear the tension or the moment; the continuum can bear the pressure, the tension and the moment. **Controlling the integrity of the subgrade can provide the basement of the system of the reaction equilibrium to resist the lateral slip of the subgrade**; (2) The soft soil foundation of the soft soil subgrade has the characteristics of a large void ratio, high compressibility and so on. The phenomenon of a lateral slip of the soft soil subgrade is easy to cause. **It is very important to control the reaction torque of the system of the reaction equilibrium of the soft soil foundation.**

In fact, the existing method of analysis and treatment of the lateral slip of the soft soil subgrade ignores the control of the system of the reaction equilibrium of the soft soil subgrade, especially when the characteristics of the soft soil subgrade near mountains and water are relatively poor. It is not surprising that the problem of a lateral slip of the soft soil subgrade is more prominent (Fig. 5.50a).

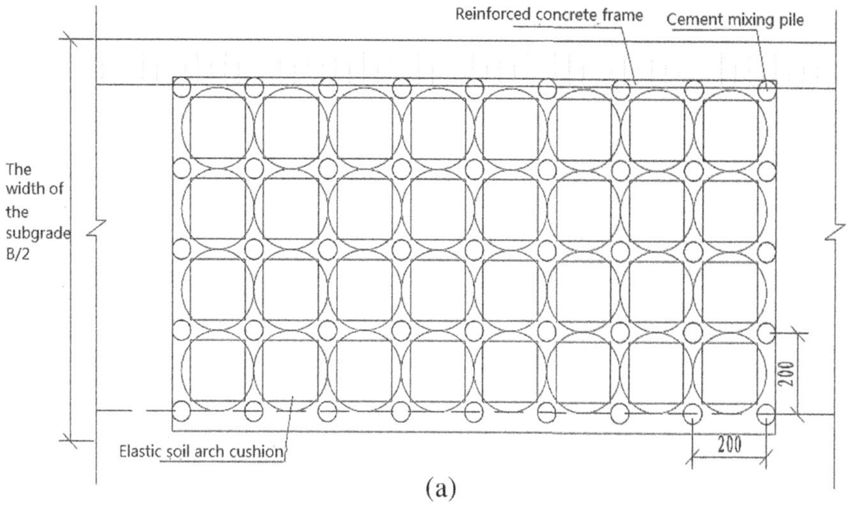


(a)

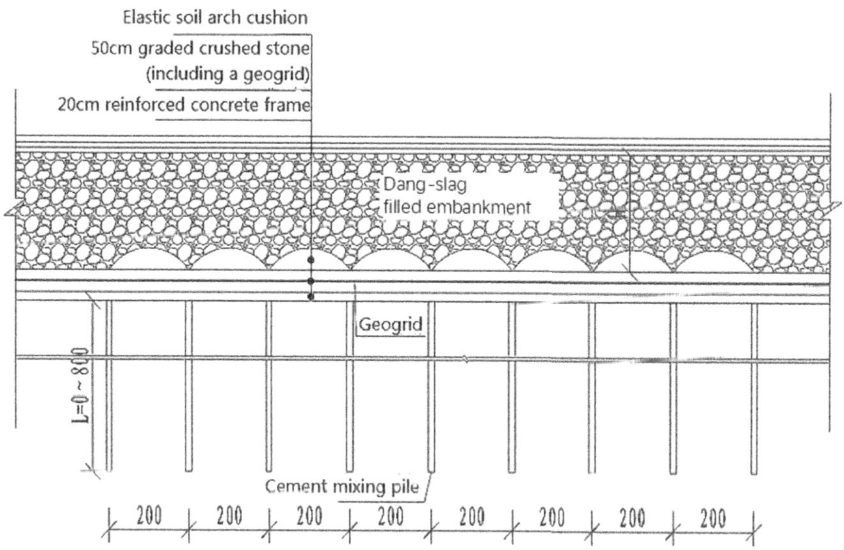


(b)

**Fig. 5.49** **a** Layout plan of the treatment of the soft soil subgrade; **b** Profile diagram of the treatment of the soft soil subgrade; (c) Elevation of the elastic soil arch cushion of the soft soil subgrade

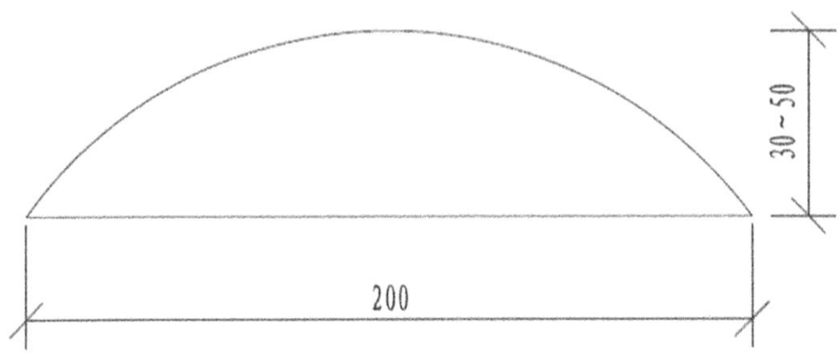


(c)

**Fig. 5.49** (continued)

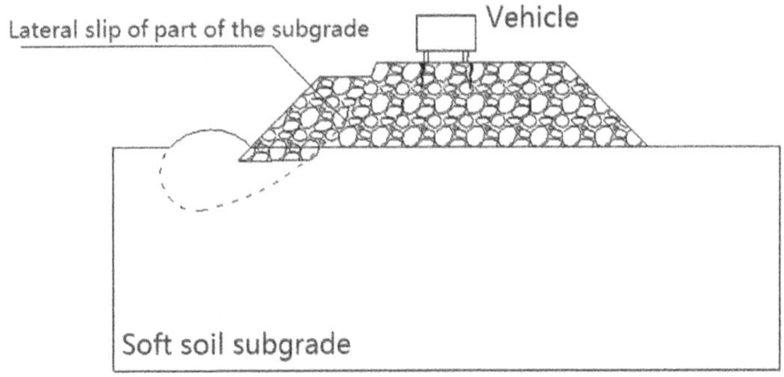


(a)

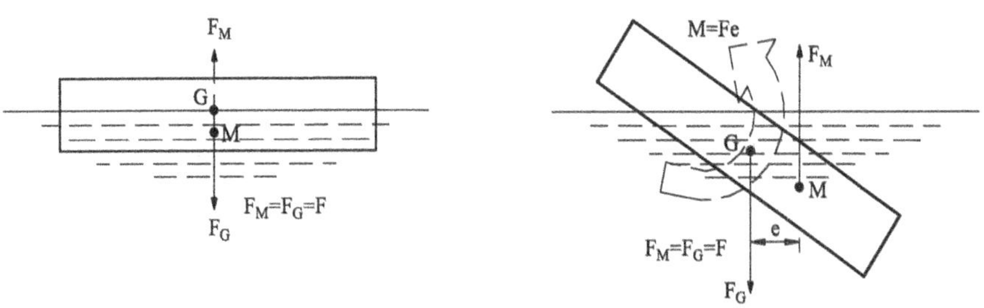


(b)

**Fig. 5.50** **a** Lateral slip of the soft soil subgrade; **b** Diagram of the analysis of the stress of the reaction torque treatment measures for controlling the system of the reaction equilibrium of the soft soil foundation; **c** Structural drawing of the bottom plate or frame and the lateral resistance piles that control the lateral slip of the soft soil subgrade on the mountain side; **d** Structural drawing of the bottom plate or frame and the lateral resistance piles that control the lateral slip of the soft soil subgrade beside the water

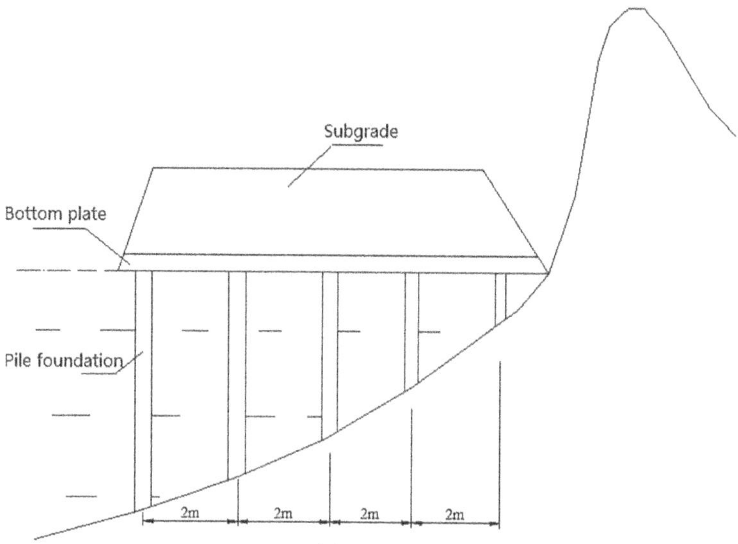


(c)

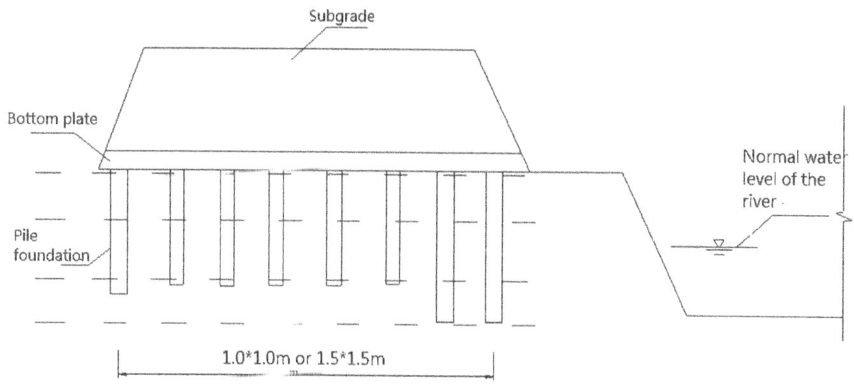


(d)

**Fig. 5.50** (continued)

**Solution** In view of the two problems that occur in the existing methods of analysis and treatment of the soft soil subgrade, the traditional theory of analysis of the soft soil subgrade ignores the control of the system of the reaction equilibrium of the soft soil subgrade. Only by simply controlling the integrity of the subgrade and providing the reaction torque can the lateral slip of the soft soil subgrade be resisted. On the basis of summarizing experience and dialectical thinking, the design of the soft soil subgrade is improved with the aim of controlling the integrity of the soft soil subgrade and providing the reaction torque.

The equation of the system of equilibrium and that of the reaction equilibrium for monolithic or bottom plate subgrade is as follows:

$$F_G = F_M \tag{5.3.1}$$

$$M_F = F_M * e \tag{5.3.2}$$

where $F_G$ is the subgrade weight; $F_M$ is the subgrade reaction force; $M_F$ is the subgrade reaction torque; e is the theoretical reaction torque of the system of the subgrade reaction equilibrium.

**Therefore, formulas** (5.3.1) **and** (5.3.2) **and Fig.** 5.50**b show that according to the coordinated control method of structural deformation, the method of designing the soft soil subgrade has improved.** In the design, the integrity of the state of the subgrade stress and deformation and the lateral resistance pile are controlled. The key technology can be adopted to control the bottom plate or frame and the lateral resistance pile of the subgrade post-construction settlement and construct the system of the reaction equilibrium of the soft soil subgrade, **so as to control the lateral slip of the soft soil subgrade.**

**Problem 5.3.3**

There are many vehicle bumps at the bridgeheads of the soft soil subgrade of the operational highways in the coastal areas of China. At present, the traffic flow of the operational highways is large and the traffic cannot be interrupted. First, excavate the original subgrade in half and then replace it with light materials to reduce the weight of the subgrade and control the subgrade subsidence, so as to solve the problem of vehicle bumps at the bridgeheads of the operational highway the soft soil subgrade.

**Solution** In view of the existing solution to the "bumps at the bridgeheads" of the soft soil subgrade of the operating highways, it is easy to ignore the relationship between the weight of the subgrade and the relative stability of the soft soil subgrade, and there is no macro quantitative control method.

Step 1 Calculate the weight of the half width part of the original subgrade, as shown in Fig. 5.51;

The width of the half width subgrade is 3.75 m, the length of a 55t vehicle is about 10 m, the excavation depth of the original subgrade is 1.2–1.5 m,

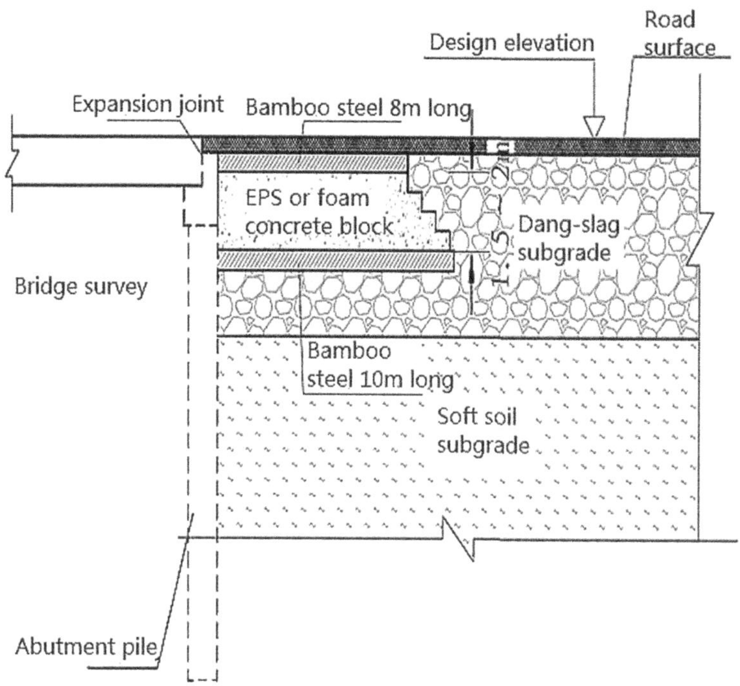


**Fig. 5.51** Diagram of using light materials to treat the "bumps at the bridgeheads" of the soft soil subgrade of operational highways

the proportion of the soil-rock subgrade is 2.0t/m$^3$, so the total weight of the original half width subgrade is 3.75 * 10 * (1.2–1.5) * 2.0 = 90–112.5t.

Step 2 Calculate the weight of the replacement light material, as shown in Fig. 5.51; The width of the half width subgrade is 3.75 m, the length of a 55t vehicle is about 10 m, the excavation depth of the original subgrade is 1.2–1.5 m, the comprehensive proportion of light materials is about 0.8t/m$^3$, so the total weight of original half width subgrade is 3.75 * 10 * (1.2–1.5) * 0.65 = 29.3–36.6t.

Step 3 Calculate the difference in the weight of the two, as shown in Fig. 5.51; (90–112.5)–(29.3–36.6) = (60.7–75.9)t

The ratio of weight difference between the two vehicles is (60.7–75.9)/55 = 1.10–1.38

By comparison, it is appropriate to excavate the original subgrade 1.5 m deep, with a margin of about 1.38. The side of the carriageway is simply supported, and vehicles can pass over the other half of the subgrade.

Step 4 Take safety measures and marks for the excavation of the half width subgrade;

Step 5 Excavate the half width subgrade 1.5 m deep and 10 m long, and excavate the side of the subgrade for the supporting carriageway;

Step 6 Level off and excavate the half subgrade, lay a 2 cm-thick and 10 m-long bamboo steel plate to control the deformation and average pressure of the subgrade, as shown in Fig. 5.51;

Step 7 Pave with EPS and other light materials on the top of the bamboo steel plate, and then pave 2 cm-thick 8 m-long bamboo steel plate to control the deformation and the average pressure at the bottom of the road, as shown in Fig. 5.51;

Step 8 Pave the road on the bamboo steel plate according to the specification requirements, and open half of the subgrade traffic, as shown in Fig. 5.51;

Step 9 Complete the other half of the subgrade according to Steps 4–8, and open all subgrade traffic, as shown in Fig. 5.51;

**Problem 5.3.4**

There are many vehicle bumps at the bridgeheads of the soft soil subgrade of the operational Highways in the coastal areas of China. At present, the traffic flow on the operational highways is large and the traffic cannot be interrupted. First, excavate the original subgrade in half and then use the lower partition to control the subgrade sinking, so as to solve the problem of vehicle bumps at the bridgeheads of the soft soil subgrade of the operational highway.

**Solution** In view of the existing solution to the "bumps at the bridgeheads" of the soft soil subgrade of the operating highways, it is easy to ignore the relationship between the settlement of the soft soil subgrade controlled by the lower partition and the relative stability of the soft soil subgrade, and there is no macro qualitative control method.

The first step is to take safety measures and marks for the excavation of the half width subgrade;

The second step is to excavate half of the subgrade 1.5–2.0 m deep and 20–30 m long, and excavate the side of the subgrade for the supporting carriageway;

In the third step, two rows of $\varphi$ 60 cm-long bored piles with a longitudinal spacing of 100 cm and a transverse spacing of 150 cm, and two rows of $\varphi$ 60 cm-long bored piles with a longitudinal spacing of 100 cm and a transverse spacing of 150 cm, and a length of 5 m, are driven near the bridgehead, as shown in Fig. 5.52;

The fourth step is to level off and excavate half of the subgrade, lay a 20 –30 m-long reinforced concrete slab with 20 cm thickness, to control the deformation and average pressure of the subgrade, as shown in Fig. 5.52;

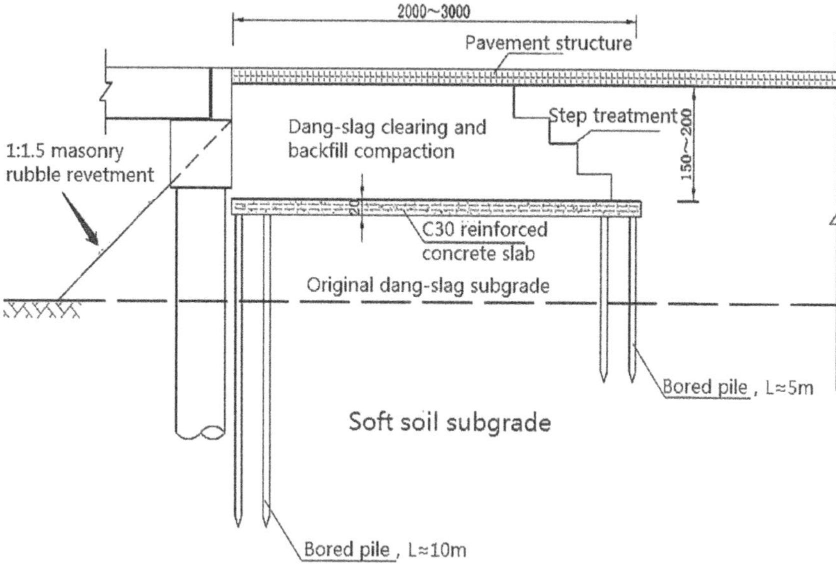


**Fig. 5.52** Diagram of treating the lower partition to avoid the "bump at the bridgehead" of the soft soil subgrade of the operating highways

The fifth step is to backfill the steps of the slag removal subgrade on the reinforced concrete slab according to the specification requirements, then make the pavement, and open half of the subgrade traffic, as shown in Fig. 5.52,

The sixth step is to make another half of subgrade according to the first step to the fifth step, and open up to all of the subgrade traffic, as shown in Fig. 5.52.

**Problem 5.3.5**

It deals with the application of an improved soil subgrade of highway—2007, the longitudinal second road bridge of the Nanjiao River of national highway 320 near Majiabang, Jiaxing, a class I highway. The project involves 10 connections (ramp), and about 120,000 cubic meters of lime soil subgrade filling. According to the design requirements of different sections, 4, 6 and 8% lime soil are respectively used for filling in the construction. 4% lime soil is used for filling between the bottom of the road trough and the cushion, as well as the subgrade and base of the sidewalk. The maximum compaction thickness of each layer is not more than 20 cm; 6% lime soil is used for the cushion layer constructed after 30 cm excavation of the mixed pile top of the soft foundation, which is divided into two layers; 8% lime soil is located within 80 cm of the road trench in the filling section, which is divided into five layers, each layer is 16 cm compacted. The subgrade should be constructed from

**Table 5.4** Relationship between CBR and ash content of the improved soil

| Ash ratio | CBR |
|---|---|
| 3% cement + 3% lime | 28 |
| 4% cement + 4% lime | 34 |
| 5% cement + 5% lime | 42 |

May 2006 to May 2007 to avoid construction in adverse seasons and rainy seasons. The temperature should not be lower than 5 °C and the construction should be stopped one month before freezing. Prevent the surface layer of lime soil from freezing. The 7-year application effect of this section of lime soil subgrade is obviously better than that of the rest of the same highway.

**Solution** The soil in the Zhangjiagang is similar to that of the Qiantang River, but slightly more viscous. Jiangsu's method is to put the materials in place, add a certain proportion of 5–8% lime to the current soil to stabilize the soil, and add 12% lime to use as the pavement sub base to replace the cement stabilized macadam sub base, and the quality is good.

The subgrade soil of the S203 provincial highway reconstruction project is mainly clay and silt (soil from the Qiantang River hydraulic fill). The quality of the soil is improved through the quicklime mixing test (5, 6, 8, 10%) of the three road sections, and the improvement performance is checked to see whether it can meet the specific index requirements in the Code through the field inspection of three indexes, namely, compactness, CBR and deflection. By means of the indoor and outdoor tests of the method for lime cement comprehensive stabilization, the consolidation effect, strength and water stability have been significantly improved, as shown in Table 5.4.

**Problem 5.3.6**

The "bump at the bridgehead" of the operating highway refers to the phenomenon of bumping and jumping when the vehicle passes across the junction of the road with the bridge rapidly due to the post-construction settlement difference between the bridge, culvert and other structures and the embankment at the back of the abutment or large longitudinal slope mutation in the engineering of the highway. At present, the soft soil subgrade is usually grouted and reinforced, and the upper part is added with the approach slab of the abutment, but the key link affecting the phenomenon of "bump at the bridgehead" is not indicated. The results show that there is no clear and reasonable bearing system for the structure, only the root state of the slurry vein is densified in the middle of the subgrade or the soft soil foundation.

**Solution** In view of the existing technology for the treatment of "bumps at the bridgeheads" for the operating highways, it is not specified that the integrity of the state of subgrade stress deformation and the rheological index of the soft soil are the key links affecting the phenomenon of "bumps at the bridgeheads", which is mainly composed of a subgrade bearing pile and a certain thickness of the reinforcement layer formed by high-pressure grouting at the bottom of the embankment, as shown in Fig. 5.53a. The reinforcement layer of the embankment grouting has a certain strength and rigidity, which can transfer the load of the upper embankment and vehicles to the lower part, forming the system of subgrade bearing under the joint action of the reinforcement layer of the embankment grouting, the piles and the soft soil, meeting the integrity requirements of the state of the subgrade stress and deformation, effectively controlling the subsequent settlement of the subgrade, and ensuring the effect of the treatment for the "bump at the bridgehead". The stress diagram of the process bearing system is shown in Fig. 5.53b.

**Problem 5.3.7**
At present, the method for the treatment of the soft soil foundation with a large water content (including beaches, dredged silt distribution areas, etc.) usually adopts a method of surcharge or vacuum preloading drainage consolidation, or a method of surcharge combined with vacuum preloading drainage consolidation, and then uses a cement mixture pile or a solidified agent mixture to form a composite foundation.

There are some limitations in the consolidation method of surcharge or vacuum preloading. The main performances are as follows: (1) The existing technology is a kind of technology for deep treatment from the surface to the underground, and the economic depth for treatment is generally not more than 10 m; (2) The construction period is long; (3) The follow-up soil will still have a slow change in the long term,

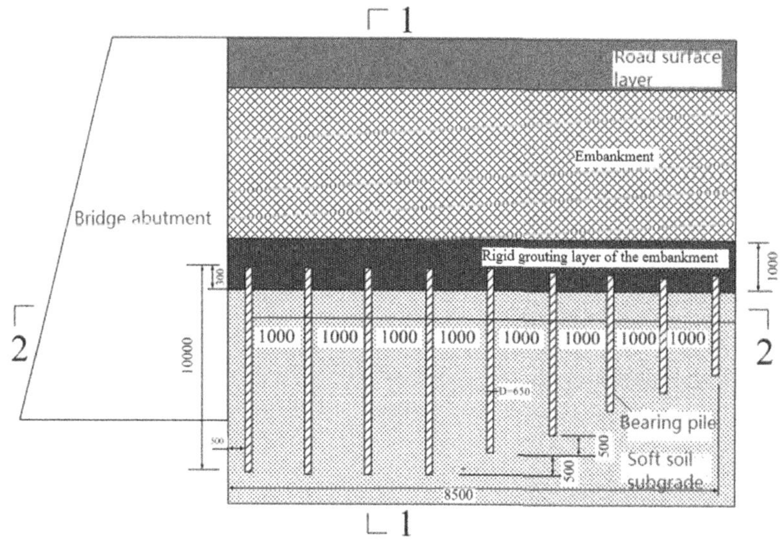


**Fig. 5.53** **a** Longitudinal structure of the bearing system. **b** Diagram of the stress of the bearing system

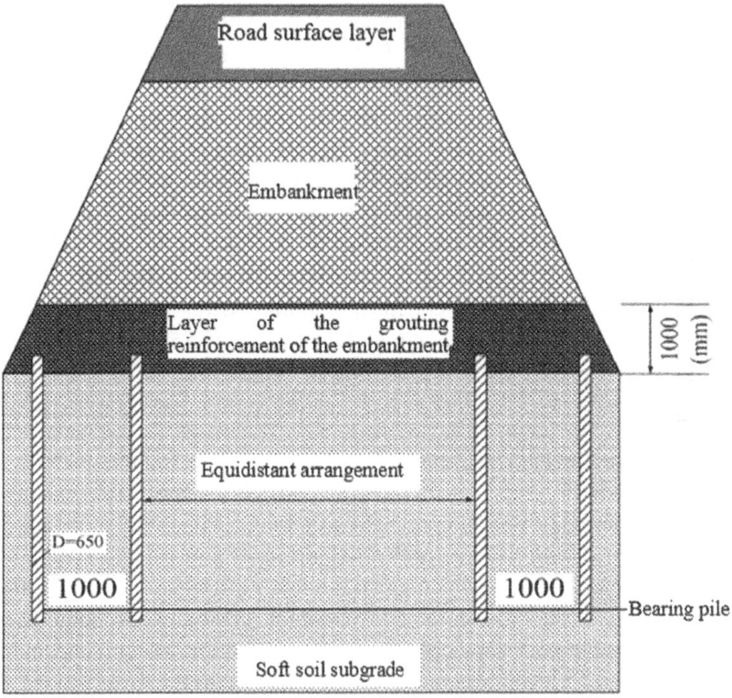

1-1Horizontal structural drawing of the soft soil system

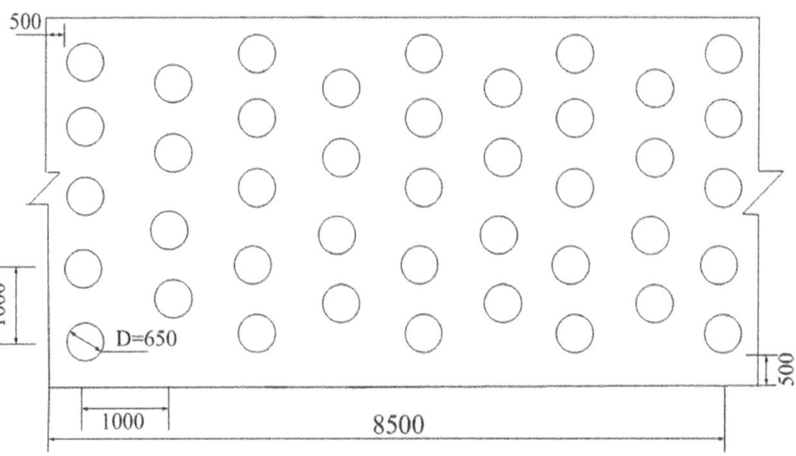

2-2 Layout plan of the bearing piles

**Fig. 5.53** (continued)

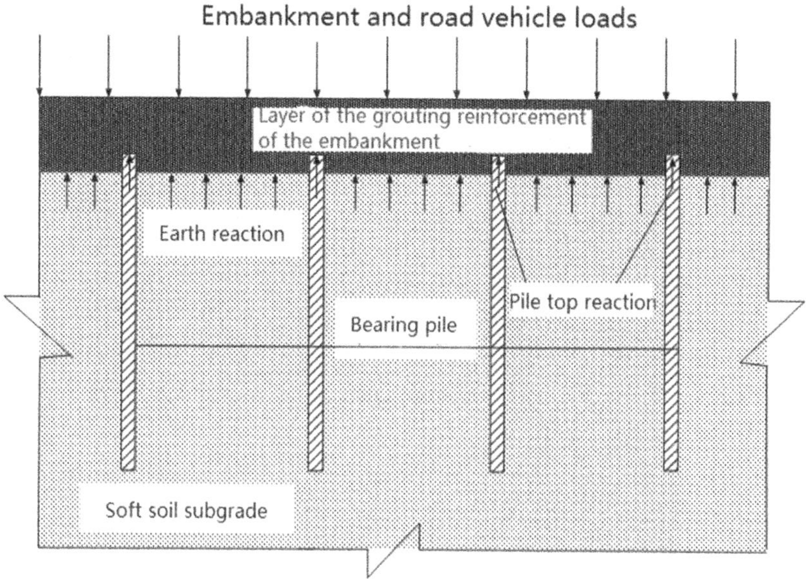


**Fig. 5.53** (continued)

the bearing capacity of the foundation is limited, and the settlement after construction is large; (4) The cost is high.

In Patent No. 201510584221.2, Wu Huiming and others developed a method of treatment via soil hardening. The method uses air injection equipment to inject air into the soil through a conduit. After the air injection is completed, the pumping operation is carried out by means of pumping equipment. The invention only improves the mechanical engineering properties of the soil, without forming a composite foundation, and the bearing capacity of the treated foundation is limited.

In Patent No. 201510301241.4, Wu Huiming and others developed a method of treatment of the soft soil foundation of different depth disturbances combined with drainage consolidation; in Patent No. 201711363451.1, Wu Huiming developed a vertical and horizontal three-dimensional layered pressurized drainage method of the treatment of foundations; both of the above-mentioned inventions are combined with surcharge loading or vacuum preloading to accelerate drainage consolidation, but the bearing capacity of the foundation after treatment is limited.

In view of the above problems, a kind of structure and method for the treatment of the soft soil foundation needs to be developed which can shorten the construction period, is cost-effective and improves the bearing capacity of the foundation, and it controls the settlement of the foundation.

**Solution** In view of the problems existing in the treatment methods of the soft soil foundation with a large water content (including beaches, dredging silt distribution areas, etc.), a method of accelerating the drainage consolidation of the soft soil foundation by means of a high-pressure air jet in a PVC pipe reserved hole, then pumping water upward, and finally forming a composite foundation by a cement mixture pile treatment is proposed (Fig. 5.54).

(1) Unlike the traditional method of preloading or vacuum preloading drainage consolidation, a PVC pipe is used first, and holes are reserved in the PVC pipe.

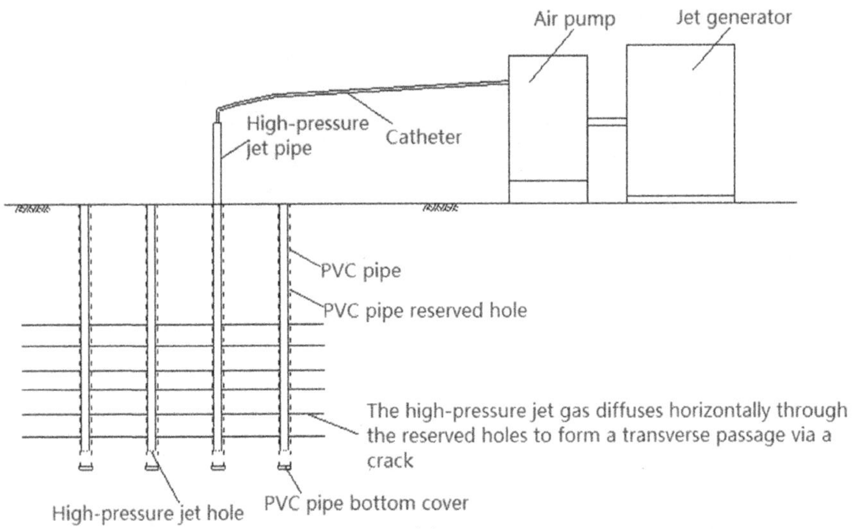


(a)

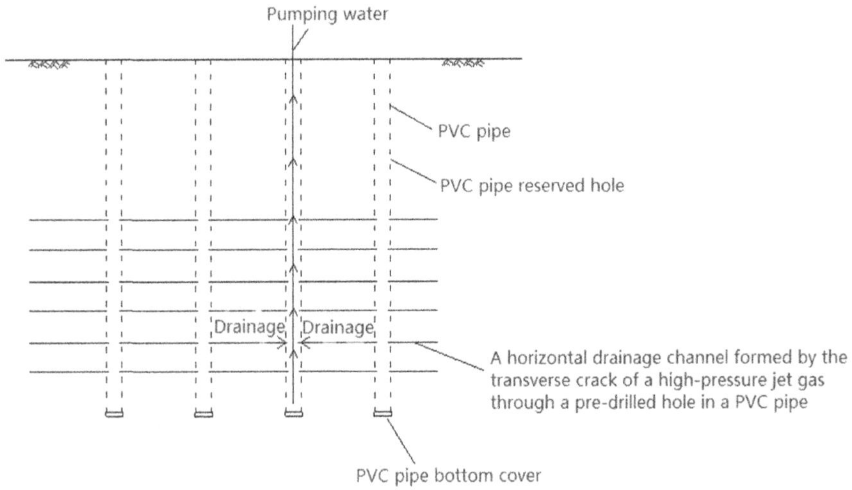


(b)

**Fig. 5.54** **a** Diagram of the high pressure jet-accelerated drainage consolidation treatment of the soft soil foundation in PVC pipes; **b** Diagram of a horizontal channel drainage formed by high-pressure air jet diffusion in a PVC pipe reserved hole and water pumping in a PVC pipe; **c** Diagram of the composite foundation formed by cement mixture piles after pumping and drainage

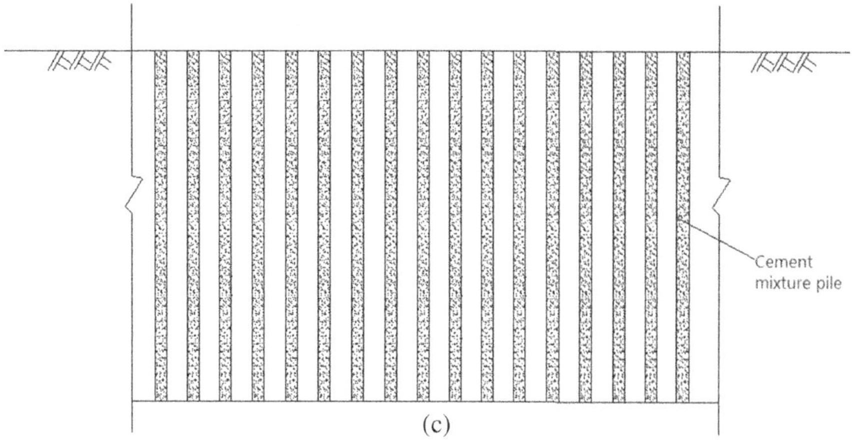


(c)

**Fig. 5.54** (continued)

A high-pressure rotary jet is used in the PVC pipe. High-pressure gas diffuses horizontally through the holes reserved in the PVC pipe to form a split transverse channel. The water in the soft soil silt converges into the PVC pipe, accelerates the drainage consolidation, and then is pumped up. This method can shorten the working period and reduce the cost.

(2) Based on the drainage mentioned in step (1), the $c$ of the cohesion and internal friction angle of the physical soil and $\varphi$ are improved, which is conducive to the construction of a cement mixture pile, that is to say, the water cement ratio is reduced, via the cement mixture pile it is easy to form early strength cement soil reinforcement, so as to improve the effect of the foundation treatment. The cement mixture pile is used to form a composite foundation to realize a rapid treatment of the soft soil foundation and improve the bearing capacity.

This method can accelerate the drainage and shorten the construction period; the treatment of the soil layer is uniform, the rapid bearing can effectively reduce the settlement of the foundation after construction; it can also save the cost and so on, which is a better method for the treatment of the soft soil foundation.

**Problem 5.3.8**

The administrative areas of our country are Erguna City, Genhe City and the Elunchun Autonomous Banner to the north of Hulunbuir City, Inner Mongolia, some parts of Arxan city in Hinggan League, Mohe County and Tahe County in the Great Khingan Mountains area of Heilongjiang Province, which are also the locations of the Great Khingan Mountains in China. The landform mainly consists of low and medium-high mountain hills and valley terraces. The annual average temperature in this area is –3 and –7 °C. The temperature in July is the highest, with an average of 17.9–19.8 °C. The highest temperature is 37.1 °C, and the lowest extreme temperature is –46 and –49 °C. There is a large area of an island-shaped permafrost in this area, which can mostly be found in the valley terrace with relatively low-lying landforms, dense vegetation (shrubs and Tatou grass), shady sloping fingers of hills and other areas. This kind of area is the high-quality corridor belt for the selection and development of highway routes, which makes it possible for the highway to cross the island-shaped permafrost area in a large range. The general depth of the upper limit of the permafrost is 0.8–7.8 m, and the general depth of the lower limit is 3.6–22.8 m. The volume of ice content is generally 0.05–0.5%, which is mainly composed of less ice-saturated frozen soil and even a layer of icy soil. For example, national highway 332 (the original provincial highway 301) from Alihe to Genhe, Genhe to Erguna, and the highway from Arxan pass to Nianzishan of S308, as well as a large number of class II, III and forest roads can be found throughout this area. A large range of highways are severely affected by island-shaped permafrost, such as a large area of uneven settlement of subgrade and pavement, with the maximum settlement depth of more than 1.0 m, which has been used as road construction. The fracture of culverts and passages of a shallow foundation often occurs, which seriously affects the normal traffic of vehicles.

At present, after years of experimentation and practice, many attempts have been made, such as increasing the thickness of the subgrade, excavating the upper shallow layer ($\leq$ 3.0 m) of permafrost, adopting the principle of half excavation and half protection, and not studying the type of permafrost, directly adopting the principle of protection. The problems of highways crossing the island-shaped permafrost have not been effectively solved and controlled, and there are still a large range of problems such as uneven settlement and the cracking of the subgrade and pavement. Therefore, the technical problems of highway construction in the island-shaped permafrost area are still under the direction of joint efforts of the highway participants in the future, and there is still a long way to go to solve them.

**Solution** In view of the advantages and disadvantages of the methods of prevention and treatment of the island-shaped permafrost problem of new highway subgrade in high altitudes and cold areas, the principal focus is on: ① Strengthening the waterproof and drainage design; ② The following measures can be taken to avoid or reduce the problems regarding the road subgrade in terms of the methods of prevention and control of the problems regarding the island-shaped permafrost of the new road subgrade in high altitudes and cold areas.

(1) Strengthen the design of waterproofing and drainage elements, eliminate the frost damage (i.e. water damage), properly reduce the groundwater level and improve the water environment within the subgrade, strengthen the drainage protection, and eliminate the concept of "heavy protection, light drainage" commonly existing in the previous project for the treatment of permafrost. Innovatively put forward the principle of "first drainage and then protection" to treat the island-shaped frozen soil, and apply the concept of the blind ditch commonly used in cutting drainage and salivary ice treatment to the treatment of the island-shaped frozen soil, adopt the gravel blind ditch and the catchment area upstream to set a water retaining ridge to strengthen the drainage protection of the frozen

soil section, so as to achieve the purpose of reducing the problems caused by frozen soil, as shown in Fig. 5.55a.

(2) Avoid disturbance to the environment or control the original ecological environment to the greatest extent (principle of frozen soil protection)
Based on the detailed analysis of the type of the permafrost and the depth of its lower limit, combined with the results of and the experience gained from previous research, for the unstable (III)—basically stable (II) frozen soil with icy soil, full ice and rich ice with a depth of the lower limit of > 5.0 m, the hot rod method is adopted to avoid the melting of frozen soil, as shown in Fig. 5.55b. For the basic stable-stable ice layer, saturated ice, rich ice and multi frozen soil with a depth of the lower limit of > 5.0 m, adopt the ventilation subgrade to avoid the melting of the frozen soil, as shown in Fig. 5.55c.

(3) For the lower limit of the frozen soil $\leq$ 5.0 m, combined with the previous engineering practice, it is proposed to thoroughly remove the island-shaped permafrost, replace it with rubble and natural gravel, and completely eliminate the problems of frozen soil, as shown in Fig. 5.55d.

(4) For the unstable (III) ice layer with soil and ice-saturated permafrost with a lower limit of permafrost > 10.0 m, according to the relevant experience of other projects, and according to the requirements of mechanical calculations and highway specifications, the low bench bridge is adopted to cross the island-shaped frozen soil or set a reinforced concrete frame at the bottom of the road to control the integrity of the subgrade, as shown in Fig. 5.55e.

## 5.4 System for Slope (Subgrade) Equilibrium and Drainage (Water Damage) Problems

The harm of water to engineering structures, such as slopes, is that the internal water of a rock and soil mass makes its cohesive force decrease and increases its sliding force; the scour force or impact force produced by the water potential of external water changes the shape or structure of the rock and soil mass and directly affects the safety of the slope retaining structure. Therefore, it is necessary to not only calculate the influence of external water and internal water on retaining structures, but to also effectively control the harm of external water and internal water on retaining structures, and the latter is more important.

I. Physical significance and method of analysis of the fluid impact of the engineering structure

There are many methods of design and construction for the problem of bridge foundation scour in mountain rivers, and the design and construction specifications also have corresponding provisions, mainly considering the flow, river width, riverbed geology and other conditions, so the design and calculation of the amount of scour is generally inaccurate. According to the impulse theorem MV = Ft, it can be seen that the amount of scour of a riverbed is not only related

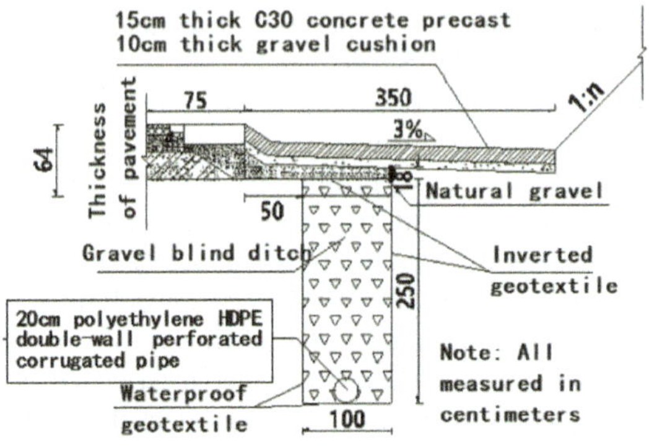


(a)

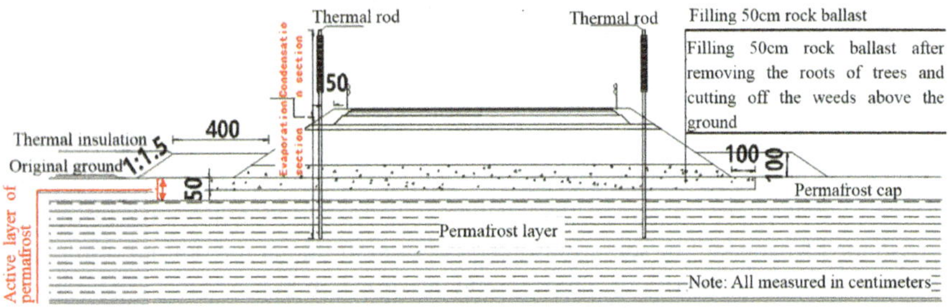


(b)

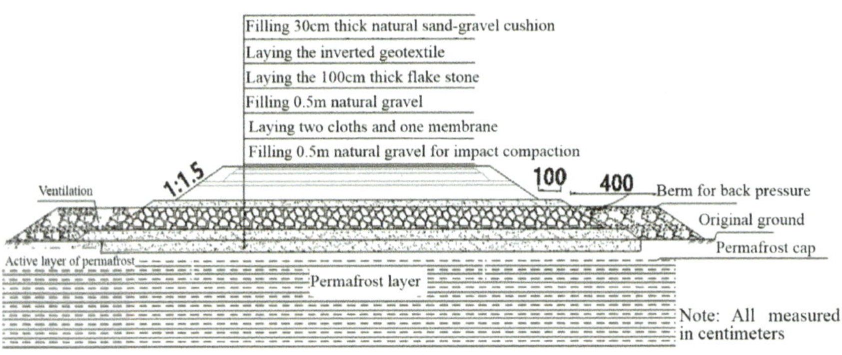


(c)

**Fig. 5.55** **a** Diagram of the gravel blind ditch used to strengthen the subgrade drainage in an island-shaped permafrost section of a new highway; **b** Diagram of the subgrade treatment of an island-shaped permafrost section by the hot rod method for a new highway; **c** Diagram of the method of ventilation subgrade applied to the treatment of an island-shaped permafrost section of a new highway; **d** Diagram of the complete removal of an island-shaped permafrost on the subgrade of a new highway; **e** Diagram of the low bench bridge of a new highway crossing the island-shaped frozen soil or subgrade with the bottom equipped with a reinforced concrete frame to control the integrity of the subgrade

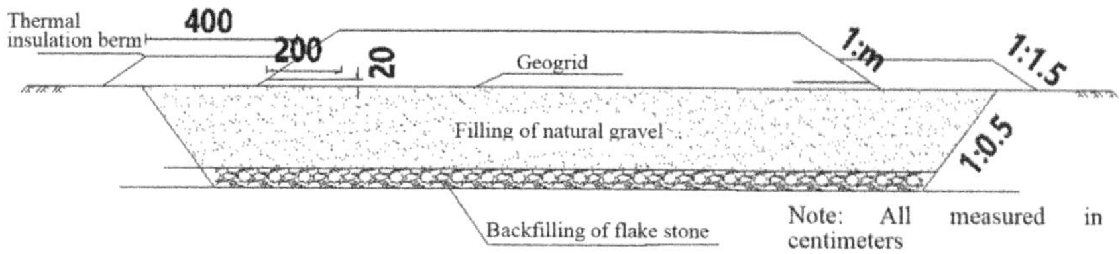


(d)

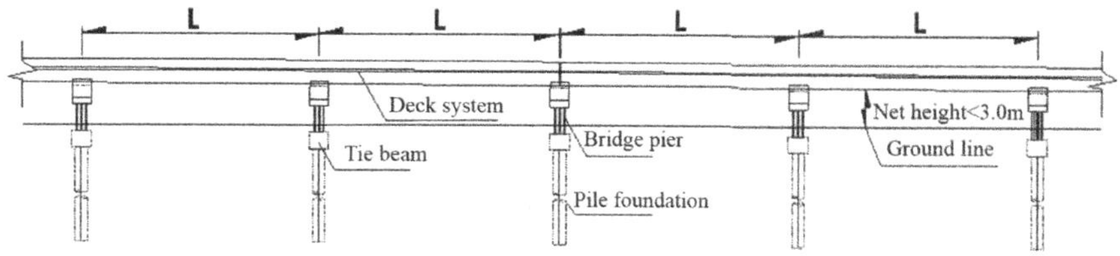


(e)

**Fig. 5.55** (continued)

to the flow, the width of the river, the geology of the riverbed and other conditions, but it is also related to the flow velocity and scour time. Among them, the geological conditions of the riverbed cannot be changed, the flow is conditionally controllable, and the flow velocity and scour time are controllable. Therefore, it is necessary to study the methods of design and construction to prevent and control the erosion of the foundation of the bridges in mountain rivers from the perspective of mechanical analysis.

In view of the existing problems of the scour of bridge foundations in mountain rivers, how to prevent and control the risk of the scour of bridge foundations in mountain rivers is not fully explained by the mechanical concept. According to the theorem of the amount of scour, MV = Ft, energy dissipation and a speed reduction surge tank is used to change the flow rate, scour time and other conditions, so as to achieve the purpose of preventing and controlling the scour of the bridge foundations in mountain rivers.

The first step is to find out the key factors that affect the scource of bridge foundations in mountainous areas according to the theorem of impulse, MV = Ft; According to the impulse theorem MV = Ft, the scour formula of the river per unit length can be established:

$$Q \times \gamma \times V = F \times t \tag{5.4.1}$$

where Q is the flow of the river per unit length; $\gamma$ is the volume weight of water; V is the rate of the flow of the river; F is the riverbed scouring force; t is the scouring time of the flow of the river per unit length.

In formula (5.4.1), the discharge Q is constant; the velocity V is controllable; the scour time t is time-delay, and the scour force F of the riverbed can be changed; while the discharge, riverbed geology and other conditions cannot be changed.

That is,

$$F = (Q \times \gamma \times CV)/t \quad (5.4.2)$$

The following can be seen from formula (5.4.2): Engineering measures can reduce the velocity V, extend the scouring time t, and reduce the scouring force F of the riverbed.

In the second step, for the scour of bridge foundations in mountainous areas, the velocity V can be reduced and the scour force F can be reduced by constructing an energy dissipation and speed reduction surge tank, and the scour time t can be prolonged;

It can be seen from Fig. 5.56 that 1–2 rows of 10–20 m long prefabricated reinforced concrete piles are driven into the riverbed at a suitable position 10–20 m downstream of the bridge foundation as the fixed piles of the dam. When a 1 $m^3$ precast concrete block is placed upstream or in the middle of the fixed pile, 1 m higher than the riverbed, and a 1 m high river dam is constructed, as a river energy dissipation and deceleration pool, the velocity $V(V_0 \rightarrow V_1)$ can be reduced, and the scouring time $t(t \rightarrow t + \Delta t)$ can be prolonged, so the scouring force $F(F_0 \rightarrow F_1)$ can be reduced. And making the river bed near the bridge foundation back silting, and gradually restore the buried depth of the bridge foundation, so as to achieve the purpose of preventing and controlling the scour of the bridge foundation in mountain rivers.

Each disaster should be distinct from others, and harm brought by the disasters (involving the safety of people and property). Sometimes, disasters are large (very) and harm is small (little). Sometimes, disasters are small (little) and harm is large (much). Sometimes, disasters are large (very) and harm is large (much). There is a system of stable equilibrium or system of a limit of equilibrium in

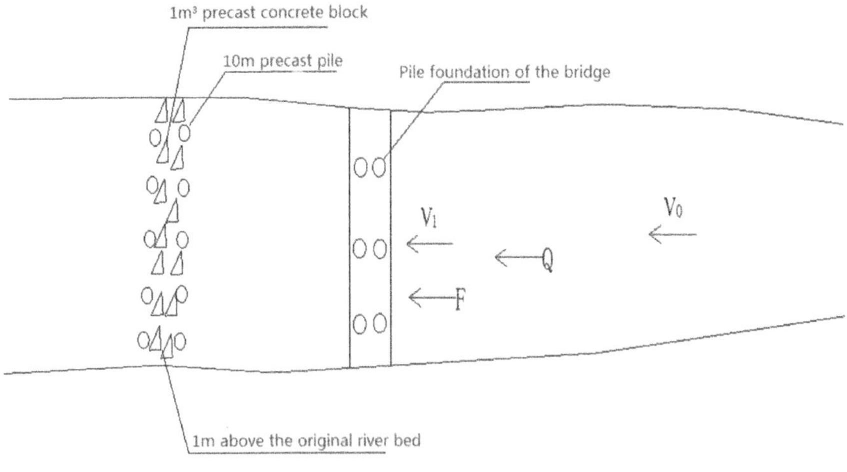


**Fig. 5.56** Diagram of river bridge foundation scour and protection

a natural slope, so the excavation slope of engineering design and construction must meet the conditions of the system of stable equilibrium, otherwise, there would be an engineering risk.

II. Treatment of the 102 landslide project of the Sichuan-Tibet Highway

The 102 Landslide is located in Yigong Township, Bomi County. When the Sichuan-Tibet Highway was built in the early 1950s, the landslide had already been a disaster precursor. In 1986, the rainfall in this area was very abundant, which caused a creeping deformation of the whole slope. On June 16, 1991, the subgrade of this section of the highway sank sharply for 2 m, and continued to sink for 1 m on June 17. On June 18, the subgrade slope began to collapse. By about 2:00 p.m. on June 20, the whole slope lost its equilibrium and suddenly fell rapidly. A large amount of material slid into the river. The front of the sliding body rushed directly to the other bank of Parlung Zangbo River, forming a blocking dam with a high north bank and a low south bank (Fig. 5.57).The blocking height of the north bank was 50 m, the height of the south bank was about 10 m, the average blocking height of the dam was 20 m, the river was blocked for 40 min, and the backwater was blocked for 3.0 km.

Every year during the rainy season, mudslides, landslides and rolling stones on the slope of the landslide surface occur constantly, resulting in a series of accidents; from June 1991 to December 2000, 20 car accidents occurred, killing 9 people. From July 2001 to November 2002, the landslide treatment measures were completed. At present, the landslide section is basically unblocked.

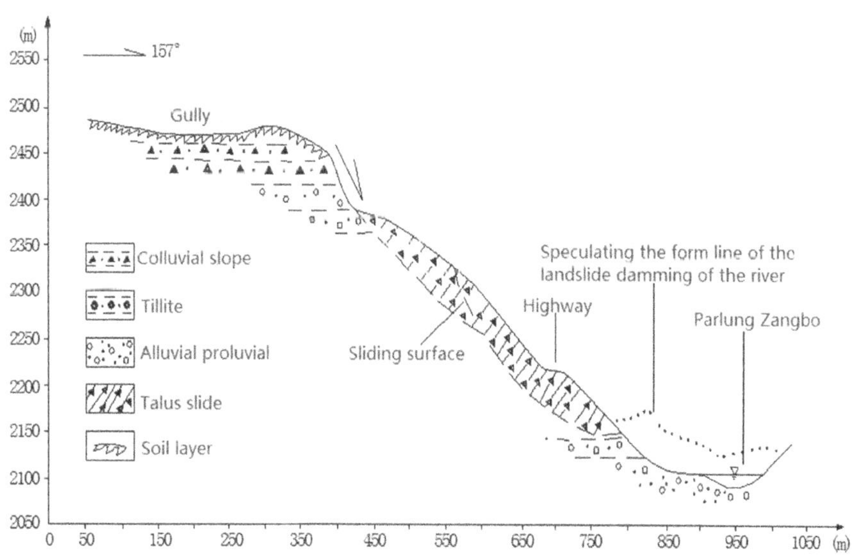


**Fig. 5.57** Cross-section diagram of the 102 landslide

The main idea of the engineering design of a landslide is "reduce, anchor, discharge and maintain", that is to say, take appropriate measures to reduce the weight and unload, increase the stability of the landslide mass, fill the gully in the middle of the landslide with the supporting type of anchor cable rib plate wall, pile plate wall, etc. to form a guarantee subgrade, with a variety of drainage facilities to eliminate surface water and underground water, so as to reduce the erosion of the landslide mass and subgrade, and further strengthen the landslide mass, the stability of the subgrade, and by strengthening the maintenance, achieve the purpose of guarantee.

The penetrating main crack in the back wall of the landslide is a big hidden danger, so it is necessary to backfill the crack tightly to prevent the surface water from penetrating into the crack. In order to reduce the erosion of the slope toe caused by the flood in Parlung Zangbo, the masonry retaining wall for gravity diversion should be built along the slope toe. Since the strength of the sliding surface decreases after sliding, anti-sliding engineering measures should be strengthened.

At the east side of the upper slope of the subgrade and the shoulder of the lower slope of the highway, 768 pre-stressed anchor cables with a diameter of 135 mm, each of which was 25–50 m long and 32870 m in total, were built (Fig. 5.58);

At present, the 102 Landslide has achieved the basic requirement of smooth flow, but the landslide has not been completely controlled. According to a field survey, Tianchi is in the mountain area at the back edge of the landslide, which is in a state of metastable equilibrium. There is often groundwater exposed, and there is often partial collapse. In the rainy season, the flow of debris on the slope is very strong, endangering the safety of the traffic along the highway.

**Fig. 5.58** Pre-stressed anchor cable rib wall for the treatment of the 102 landslide

**Fig. 5.59** The pavement of the Sichuan-Tibet Highway destroyed by the 102 landslide group collapse

Although some engineering measures have been taken to solve the problem of groundwater, such as setting a drainage ditch, etc., but the groundwater activities have not been effectively controlled, which makes the local slope collapse, slope debris flow and gully still develop, endangering the safety of driving along the road. In addition, due to the good water collecting environment and thick alluvial proluvial gravel sand with good permeability on the upper part of the slope, a large amount of groundwater enters the landslide mass, which is also unfavorable to the overall stability of the landslide (Fig. 5.59).

The 102 landslide group is described as "total repair and total deterioration, like a sore that cannot be cured by a block", which has become an unavoidable difficulty in the construction of the 102 landslide group in Tibet.

The total length of the 102 landslide group renovation project of the Sichuan-Tibet Highway is about 3.6 km, including a 1.725 km tunnel, which will be constructed according to the technical standards of class III highways. The overall design speed of the route is 30 km/h, the subgrade width is 7.5 m, the pavement width is 6.5 m, and the design load of the bridge is conformable to the standard of a highway class II; the design speed of the new tunnel is 40 km/h, the tunnel width is 9 m, and the net height is 5 m. It is absolutely necessary to completely get rid of the influence of the flow of debris associated with collapse of the slope!

III. The flow of associated debris caused by the collapse of urban soil

Satellite photos (Fig. 5.60) show that in 2005, this was the valley excavated by the quarry; in 2013, the quarry was out of service, and a large amount of water

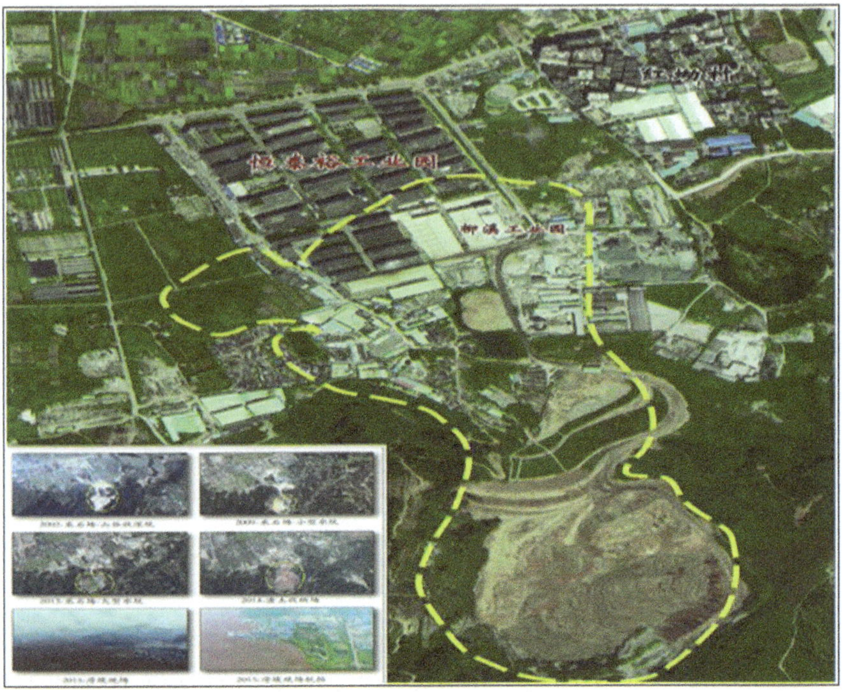


**Fig. 5.60** Impact range of the flow of the associated debris caused by the collapse of the soil

can be seen in the valley; in 2014, the abandoned quarry valley turned into a muck landfill site, and the outlet of the valley was facing the industrial park. Geological experts confirmed that the flow of debris covers 60,000 square meters, with an average thickness of about 6 m. It was found that the collapsed body in a new area of a city is a man-made pile of soil, and the original mountain body does not slide. The site of the collapse of the stacked up artificial soil belongs to the receiving site of residual mud and muck, which is mainly used for stacking muck and construction waste. Due to the large amount and steep slope of stacking, it leads to instability and collapse, resulting in the collapse of many buildings. The collapse of the mound soil induces the associated flow of debris to rapidly impact the structure of the industrial park at the exit of the valley, resulting in the collapse of many buildings and many casualties, rather than the usual meaning of the collapse of a mound soil landslide (Fig. 5.61).

IV. Method and sequence of treatment of the highway slope and valley in mountainous areas

The abnormal hydrodynamic conditions will induce the slope slide and the flow of debris (Fig. 5.62). The uplifting of the groundwater level in the deep part of the slope caused by rainfall was the main cause of the landslide (Fig. 5.63).

(a)

(b)

**Fig. 5.61** **a** Comparison of the range of influence of the flow of the associated debris caused by the collapse of the soil; **b** An actual scene of the range of influence of the flow of the associated debris caused by the collapse of the soil

**Fig. 5.62** Photo of the collapse of the slope after the rainfall

**Fig. 5.63** Photos of the collapse caused by inadequate support to the slope

The improper methods of excavation and support caused the slope collapse (Fig. 5.64).

Methods for the prevention and control of slope stability: (1) Survey the scope or surface line that directly affects the sliding of the slope and the scouring trend; (2) Remove the deep water of the slope and control the source water

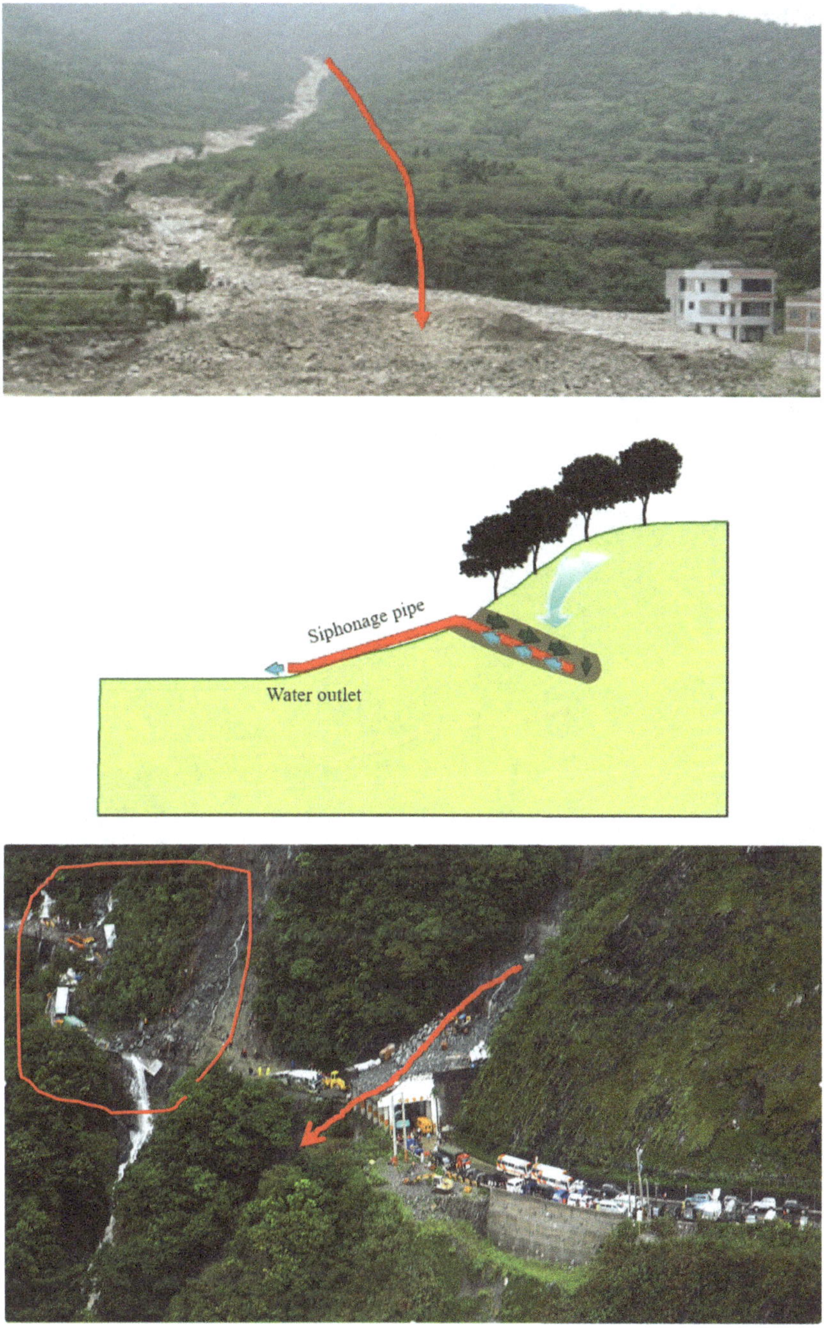


**Fig. 5.64** Comparison of setting up a tunnel or an open tunnel to prevent gully scour from damaging the highway

Fig. 5.64 (continued)

of the scouring slope (The value of c and ø directly affects the analysis of the stress, the condition for the beginning of the scouring of the source water, the drainage of the supporting layer, the surface drainage, etc.); (3) The sequence of excavation and support directly affects the basic maintenance of the original state of the slope stratum (basis of theoretical analysis, otherwise, the difference is not easy to estimate); (4) The construction process of the retaining structure is reliable, and the masonry retaining wall with poor reliability is not used (Currently, there is no good master of masonry technology) (Fig. 5.65).

**Fig. 5.65** Photo of the method of treatment for a sliding block of the tunnel portal slope

Figure 5.66 shows that the side slope with a height of 1–1.5 m on the top of a tunnel portal is cut by sliding blocks. The upper cut is about 1 m wide and 2–3 m deep. The lower cut is protruding and leaking. The length of the sliding block is about 15 m high and about 10 m wide and 6–10 m wide. The volume is about 500–1000m3. The flow of the tunnel is about 17,000 vehicles/day, 15,000 vehicles/day more than the design flow. The sliding block at the upper part of the tunnel portal has seriously threatened the safety of the traffic and personnel. It is suggested that: (1) The front and rear nodes of the tunnel should be closed to passage, and the adjacent tunnels should be organized to pass in two directions; vehicles should be organized to bypass the intersection of the highway; (2) The upper incision of the sliding block should be closed; (3) The risk assessment of the sliding block should be organized and the plan for reinforcement and reconstruction should be implemented.

Evaluation of the four reinforcement and reconstruction plans: (1) The plan of an anchor bolt strengthening the sliding block: Considering that the lower part of the sliding block has been cut out, the drilling impact of the down-hole drill may trigger the sliding of the sliding block, so there are construction safety risks, and the anchoring effect is not good; moreover, it is difficult to guarantee the quality of the anchoring project after reinforcement, so the plan is not appropriate; (2) The plan of an anti-slide pile support: Considering the distance between the lower cut of the sliding block and the lower edge of the hole, the slope is only 1–1.5 m, and there is no equilibrium in front of the anti-slide pile. The blasting construction for the anti-slide pile foundation pit may trigger the slide of the slide block, and so there are construction safety risks, and it is difficult to make sure that the anti-slide effect can be put in place, which is not appropriate; (3) Excavation of sliding blocks for unloading and reinforcement: The plan may

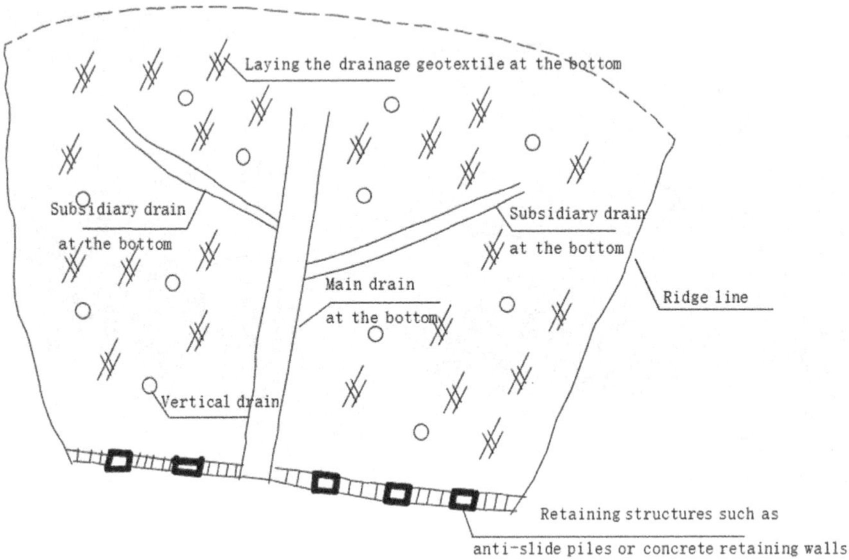


**Fig. 5.66** Diagram of the drainage and retaining structure of waste dumps in mountainous areas

affect the system of equilibrium at the top of adjacent tunnels, increase the stress on the tunnel lining, and may cause cracks in the tunnel lining, and the subsequent stability of the portal slope; in addition, the roads in mountainous areas are very narrow, and the reconstruction of roads, waste disposal areas and other links affect the ecological environment, which does not meet the requirements of a beautiful China; (4) The plan for an extension of the open tunnel: First remove the side. As the cut of the sliding block is 1–1.5 m above the top of the tunnel portal, the filling of the top of the tunnel after the open tunnel is completed is more than one time the height of the diameter of the tunnel, and then some sliding blocks are excavated and symmetrically filled on both sides of the open tunnel and on the top of the tunnel, which is not only the waste slag filling site but also the equilibrium of the remaining sliding blocks, and does not affect the ecological environment, which is beneficial; check the stability of the upper slope of the backfill of the extended open cut tunnel, and then reinforce it with anchor bolts or anchor rods.

**Problem 5.4.1**

With the reconstruction and construction of large and medium-sized cities in China and the construction of many mountain roads, many large-scale spoil areas need to be set up to ensure the stability and safety of the spoil areas, and to avoid the spoil from becoming geological disasters. Therefore, it is necessary to study and improve the design of the drainage system and the methods of construction for the prevention and control of the flow of debris in large-scale spoil areas in mountainous areas from the analysis of the drainage of the spoil areas.

**Solution** In view of the lack of a bottom permeable structure and a vertical permeable structure in the drainage of large-scale waste dumps in mountainous areas, it is easy to make the soil moisture content of large-scale waste dump reach saturation, quickly reduce the cohesive force of the soil, generate downward and outward thrust, increase the pressure of the retaining structure at the mountain pass of the waste dump, and the most unfavorable situation will destroy the retaining structure, resulting in disaster due to the flow of debris

According to the law of impulse MV = ft, the formula for a disaster due to the flow of debris $Q \times \gamma \times v = f \times t$ can be established for the large-scale waste dump. The risk of this kind of disaster in the large-scale waste dump can be directly analyzed by the mechanical method, and the method of controlling a disaster due to the flow of debris in the large-scale waste dump can be found.

The method of controlling the risk of a disaster due to the flow of debris in large-scale waste dumps is directly by a mechanical method;

It can be seen from Fig. 5.66 that the formula for a disaster due to the flow of debris of large waste dumps is

$$Q \times \gamma \times V = F \times t \tag{5.4.1}$$

where Q is the total amount of the flow of debris; $\gamma$ is the bulk density of that flow; V is the velocity of the flow of debris; F is the impact force of the flow; t is the impact time of the flow.

It can be seen from formula (5.4.1) that only by controlling the moisture content of the flow of debris Q in the soil layer of large-scale waste dumps will the cohesive force of the soil layer increase, reduce the downward and outward thrust, reduce the pressure on the retaining structure at the mountain pass of the waste dump, and ensure the stability and safety of the retaining structure; naturally, the velocity V and impact time t of flow of debris in large-scale waste dumps are controlled, so as to prevent and control a disaster due to the flow of debris in large-scale waste dumps.

(1) Clear the debris on the surface of the waste dump and repair the slope of the surface, and properly incline it to the main and branch ditches of the bottom drainage, which is conducive to the flow of the water towards the bottom to the main and branch ditches of the drainage, so that the accumulated water can be quickly discharged from the waste dump;
(2) Build main and branch ditches for bottom drainage;
Along the lowest part of the valley, the foundation of the main drainage ditch at the bottom is 1.7 m wide and 1.2 m deep; the foundation of some drainage ditches is 1.2 m wide and 0.7 m deep; then the bottom of the drainage ditch is built by mortar or the bottom of the prefabricated semicircle pipe culvert is installed, and the main drainage ditch is 1.5 m wide and 1.0 m deep; some drainage ditches are 1.0 m wide and 0.5 m deep; this is conducive to drainage without scouring the foundation of the drainage ditch. Finally, the top of the grouted drainage ditch or the top of the prefabricated semicircle pipe culvert should be installed. The main drainage ditch should be 1.5 m wide and 0.5 m

deep. Several branch drainage ditches should be 1.0 m wide and 0.5 m deep. The spray holes on both sides of the surface should be 5 cm long and 25 cm apart, which is conducive to the drainage of water at the bottom of the waste dump;

(3) A 50 cm thick gravel permeable layer is laid at the bottom of the waste dump, and a layer of geotextile is laid on it, which is conducive to the drainage of water at the bottom of the waste dump;
(4) For every 10 m × 10 m or 15 m × 15 m above the geotextile at the bottom of the waste dump, 20–0 cm PVC vertical drainage pipes with a thickness of 5–10 mm and 2 m-long PVC holes at each section are set, which is conducive to the drainage of water in the soil layer of the waste dump;
(5) According to the geological conditions of the foundation of the retaining structure at the mountain pass of the waste dump, the retaining structure, such as anti-slide piles or a concrete retaining wall, should be in accordance with the specifications;
(6) A circular intercepting ditch is set up, which is arranged along the surrounding mountains at an elevation of more than 1 m above the top of the waste dump, and extends to the natural side ditch at the outlet of the waste dump;
(7) Excavate the temporary longitudinal drainage ditch along the center of the surface of the waste dump, which is connected with the natural trench; arrange the temporary intercepting ditch along the longitudinal drainage ditch of the waste dump to the two sides of the top surface of the waste dump, which is connected with the longitudinal drainage ditch; remove the surface water in the middle of the waste dump landfill;
(8) A longitudinal drainage ditch should be set up along the center of the top surface of the waste dump and connected with the natural trench; an intercepting ditch should be arranged in the form of branches along the longitudinal drainage ditch of the waste dump to both sides of the top surface of the waste dump and connected with the longitudinal drainage ditch.

**Problem 5.4.2**

Excavation of the subgrade slope by the mountain (Fig. 5.67a) can make use of the huge advantages of local materials to fill the subgrade, but the excavation of the slope will destroy the in situ stress equilibrium. There will be shear stress in the slope, and if it is greater than the basic shear strength, the stability of the slope will be destroyed.

Because of its particularity, the excavation of the subgrade slope by the mountain is constructed rapidly. It needs to find a simple and easy-to-operate method of reinforcement treatment to fill the subgrade with local materials. At present, in order to carry out a reasonable reinforcement, a method of limit equilibrium and a method for finite elements are often used for the analysis and calculation of the stability of the slope. The method of limit equilibrium can assume the failure of the sliding surface (arc surface and broken line shape, as shown in Fig. 5.67b, c respectively) in advance, or automatically search the arc failure surface to get its safety factor quantitatively, but it cannot grasp the local failure phenomenon or displacement deformation that

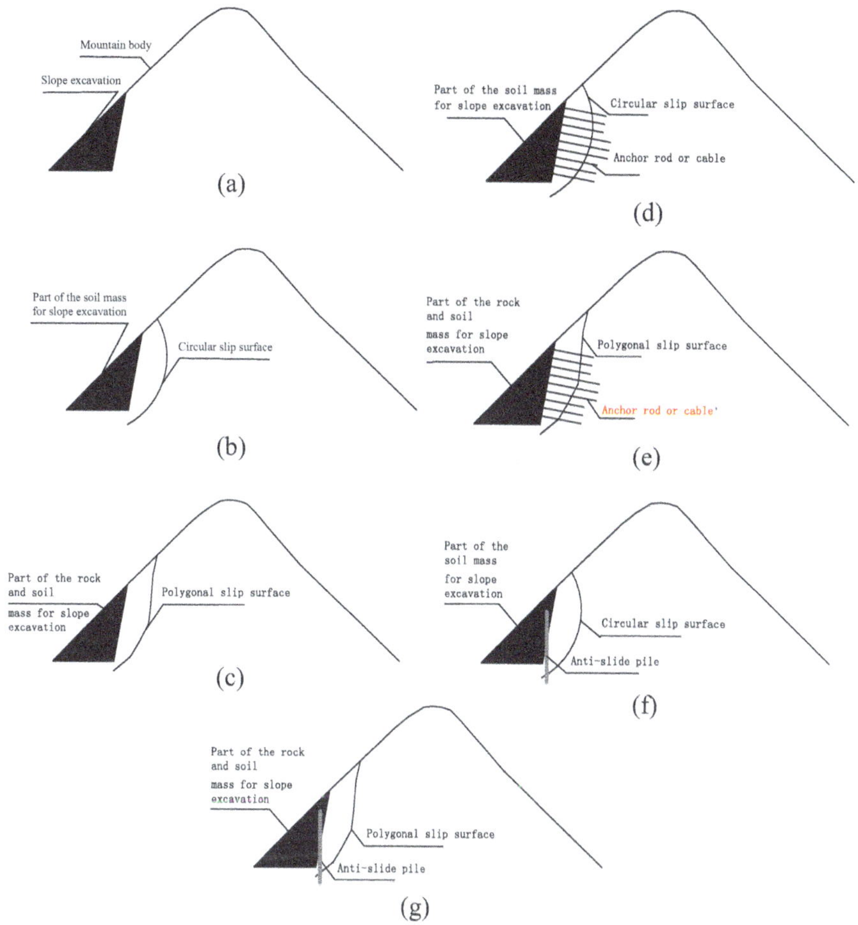


**Fig. 5.67** **a** Diagram of the excavation cross-section of a subgrade slope by the mountain; **b** Slide caused by the disruption of the in situ stress equilibrium due to the excavation of the subgrade slope (soil) by the mountain; **c** Slide caused by the disruption of the in situ stress equilibrium due to the excavation of the subgrade slope (rock and soil mass) by the mountain; **d** Diagram of the anchor bolt (cable) setting to control the landslide caused by the excavation of the subgrade slope (soil) by the mountain; **e** Diagram of the anchor bolt (cable) setting to control the landslide caused by the excavation of the subgrade slope (rock and soil mass) by the mountain; **f** Diagram of the anti-slide pile setting to control the landslide caused by the excavation of the subgrade slope (soil) by the mountain; **g** Diagram of the anti-slide pile setting to control the landslide caused by the excavation of the subgrade slope (rock and soil mass) by the mountain

may occur inside the slope; in order to master the overall mechanical trace of the slope and finally confirm the stability, the finite element numerical analysis is needed. At present, the most widely-used method is the strength reduction method to finally calculate the surface of critical failure. In the general process of construction, the soil is usually excavated first, and only after the excavation of the slope has been completed, according to the slope specification, can the assumed sliding surface of the failure obtained from the slope stability analysis and calculation be anchored or can the anti-slide pile be driven. However, the landslide often occurs in the side slope of mountain subgrade obtained by this method, which shows that the slope strengthened by this method is not very stable.

**Solution** In view of the sliding problem caused by neglecting the in situ stress equilibrium in the excavation of the existing subgrade slope by the mountain, this paper starts from the system of reaction equilibrium of the slope, and provides a method for a reaction equilibrium and stability reinforcement for the excavation and reinforcement of the subgrade slope by the mountain to control the stability of the subgrade slope by the mountain and prevent the slope from sliding.

The existing theory of analysis is used to start with the sliding surface (determine the safety factor of the sliding surface according to experience or simulate the sliding trace by means of the finite element method), and then carry out reinforcement with bolt or anchor cable through the sliding surface after excavating the soil. However, based on the applicant's findings, this method has defects. Although the method of limit equilibrium and the finite element method can determine the assumed sliding surface of the failure of the slope excavation, after the soil is excavated, the system of reaction equilibrium of the slope is destroyed, and the assumed sliding surface of the failure that did not exist in fact turns into the real potential sliding surface of the failure. At this time, the anchor bolt or anchor cable reinforcement, although theoretically able to control its sliding, but because the sliding surface already exists, it may still produce a slope collapse accident under some adverse conditions in reality.

On the basis of summing up experience and dialectical thinking, the existing method of design and construction of reinforcement and treatment of the subgrade slope by the mountain has improved from the point of view of the system of reaction equilibrium around the in situ stress equilibrium. In this method, the passive earth pressure (the most unfavorable situation of the project) of the excavated part of soil is calculated by soil mechanics, and then the passive earth pressure is evenly distributed to the friction resistance between each anchor bolt or anchor cable and the surrounding soil or the resistance of the anti-slide pile during the excavation process, so as to control the stability of the subgrade slope by the mountain and prevent the slope from sliding.

The method of reaction equilibrium and stability reinforcement for the excavation and reinforcement of the subgrade slope by the mountain is as follows: excavate the subgrade slope by the mountain in layers from top to bottom, and support the excavated part in time with an anchor bolt or anchor cable while excavating.

When excavating to the lower part of the subgrade slope by the mountain, the anti-slide pile can also be driven into the soil mass of the slope first, then the soil mass can be excavated, and the anti-slide pile can be used to replace the anchor bolt or anchor cable to support the lower part of the slope. The elastic-plastic deformation of the anti-slide pile is small, and the possibility of a potential slip surface is small. It should be noted that the upper part and the lower part do not necessarily refer to the parts above and below the central line of the slope. In the actual process of construction, they can be adjusted according to the natural and construction conditions on site.

The selection of the anchor bolt should meet the following conditions:

The standard value of the axial tension of a single bolt $N_t$ is:

$$N_t = \frac{E_p}{n \cdot \cos a}$$

where $E_p$ is the passive earth pressure of the subgrade slope to be excavated; $a$ is the angle of the anchor; n is the total number of anchor bolts set on the slope;

The design value of the axial tension of a single bolt $N_a$ is:

$$N_a = \frac{\gamma_Q \cdot E_p}{n \cdot \cos a}$$

where $\gamma_Q$ is the partial coefficient of the load;

The cross-section area of the anchor bolt $A_s$ satisfies:

$$A_s \geq \frac{\gamma_0 K N_a}{\xi_2 f_y}$$

where $\xi_2$ is the coefficient of the working condition of the anchor tension;$\gamma_0$ is the importance coefficient of the slope engineering; $f_y$ is the design value of tensile strength of the anchor bolt; K is the safety coefficient;

The anchorage length of the anchor solid and stratum $l_a$:

$$l_a \geq \frac{K N_t}{\xi_1 \pi D f_{rb}}$$

where $\xi_1$ is the coefficient of the working condition of the bond between the anchor and the stratum,$D$ is the diameter of the anchor, $f_{rb}$ is the characteristic value of the bonding strength between the stratum and the anchor;

The anchorage length between the anchor bolt and the anchorage mortar $l_b$:

$$l_b \geq \frac{\gamma_0 K N_a}{\xi_3 n \pi d f_b} = \frac{\gamma_0 \gamma_Q K N_t}{\xi_3 n \pi d f_b}$$

where, $\xi_3$ is the coefficient of the bond working condition between reinforcement and mortar; d is the diameter of the anchor bolt; $f_b$ is the design value of the bonding strength between reinforcement and anchor mortar.

The length of the anchorage section of the anchor bolt is the larger value in $l_a$, $l_b$ i.e. max ($l_a$, $l_b$); in engineering, it follows the adverse (or adverse combination) situation, and in order to ensure safety, the larger value of the two is taken. That is to say, the longer the anchor, the more stable the slope, but it cannot be infinite. This is similar to the value of the characteristic value of the vertical bearing capacity of the cement soil mixture pile in the composite foundation. The characteristic value of the bearing capacity of a single pile calculated by the strength of the pile body, the soil around the pile and the soil resistance at the end of the pile is adopted respectively. In order to ensure the safety, the unfavorable situation is taken, that is, the smaller value of the two.

The total length of the anchor bolt is the sum of the length of the anchorage section and the length of the free section of the rod.

The design of the parameters of the anchor bolt is improved according to the existing specifications. The rod under the design parameters can further ensure the stability of the slope after the excavation of the soil.

The length of the anchor bolt in the middle waist should be longer than that of those at the upper or lower end of the slope.

The length of the free section of the middle-waist bolt in the slope is determined according to the actual experience or the trace of the failure surface by the finite element analysis.

The middle-waist bolt plays a connecting role in the whole slope reinforcement system. The slip surface of the slope usually presents a circular arc and a broken line. The assumed slip surface of the waist part is a long distance from the slope surface, so its length needs to be extended so that the anchorage section can be located behind the assumed slip surface. If the length of the anchor bolt in the middle waist is not enough, it will lead to the anchoring part being placed in front of the assumed slip surface, the anchoring effect will be greatly weakened, and the force of the bolt is not enough to maintain the stability of the slope, which will eventually lead to the instability and collapse of the slope.

The anti-slide pile should be able to bear the soil thrust F:

$$F = m\frac{\gamma_Q \cdot E_p}{n}$$

where m is the number of the designed anchor bolts within the anchorage depth of the anti-slide pile.

Therefore, the anti-slide pile meeting the thrust requirements can replace the original anchor bolt and play a role in the stability of the lower half of the soil. At the same time, it makes full use of the low deformation characteristics of the anti-slide pile to keep the middle and lower soil in the state of reaction equilibrium.

The length of the free section of the anchor bolt should be determined in advance according to the relevant specifications of the slope.

The plane layout of the anti-slide pile is determined based on the comprehensive consideration of the stratum property, the magnitude of the stress, the slope of the sliding surface, the thickness above the sliding surface, construction conditions, the type of pile and the section size of the pile, the possible anchoring depth and the geological conditions of the anchoring section.

The selection of the type of anti-slide pile should be based on the comprehensive consideration of the nature of the landslide, the geological conditions at the landslide, the magnitude of the stress, the project cost, the construction conditions and the requirements of the construction period, and the engineering experience of the designer.

The traditional method of slope reinforcement is to excavate the soil before supporting, so the potential slip surface can easily become the real slip surface, and the probability of slope problems is high. When excavating the slope soil from top to bottom, support the excavated part in time while excavating; when excavating to the bottom half, the anti-slide pile can also be driven first, and then excavate the soil, so that the sliding surface will not actually occur, which greatly improves the stability of the slope. In addition, the parameters of the anchor and the anti-slide pile are optimized, which can better control the stability of the slope.

## 5.5 Similarity of the Overall Collapse of Bridges

(1) Overall collapse of a highway arch bridge in a province (conditional temporary system of equilibrium) (Fig. 5.68)

(2) Overall collapse in the demolition of a cave in a province (conditional temporary system of equilibrium) (Fig. 5.69)

(3) Overall collapse of a highway arch bridge in a province (Anji) (conditional temporary system of equilibrium) (Fig. 5.70)

(4) Overall collapse of a provincial highway bridge under overload conditions (conditional temporary system of equilibrium) (Figs. 5.61, 5.62, 5.63, 5.64, 5.65, 5.66, 5.67, 5.68, 5.69, 5.70, 5.71 and 5.72)

(5) The demolition and reconstruction of similar highway bridges in a province to avoid possible accidents due to failure (conditional temporary system of equilibrium) (Fig. 5.73)

The design load of a 5-hole unequal span ordinary reinforced concrete beam bridge on a provincial highway is conformable to the standard of highway class II. The superstructure of the bridge is a double cantilever structure with a hanging beam, the first and fifth spans are simply-supported beams, the second and fourth spans are simply-supported double cantilever beams, and the third span is a hanging beam. Due to the increasing volume of traffic and vehicle load, now the bridge is being operated under overload conditions, part of the structure has aged, and many serious problems have appeared on the bridge. There are inclined cracks near the fulcrum of the double cantilever beam. The main beam bracket and the end corner of the hanging beam have a 45° inclined crack. The displacement of the hanging beam and one end of the cantilever beam are dead, resulting in the damage and loss of the concrete at the end of the hanging beam and the crushing of the cantilever beam end. The hanging beam is likely to fall off under the external force, endangering the traffic, navigation under the bridge and the safety of the bridge itself. The bridge deck is uneven, with cracks, bulges and waves; the railings are incomplete, distorted and deformed, and the columns are damaged; the expansion joints are damaged, with broken edges, and there is the phenomenon of a vehicle bump.

(In the picture, it is shown that the proportion of piers of new arch bridges to the structures is high, while the arch ring is thin, which can easily cause the instability of the arch body and the disequilibrium of piers and arches.)

(In the picture, it is shown that a hole to the east of the newly-built arch bridge collapses first, causing the three piers in the middle of the bridge to fall to the east.)

**Fig. 5.68** Photo of the overall collapse of a provincial highway arch bridge

(6) Asphalt concrete on a bridge deck can easily be damaged due to insufficient structural rigidity of highway bridges in a province (Fig. 5.74)

In the early nineties, some bridges built by "optimized design" on national and provincial roads had insufficient structural rigidity and excessive vertical deformation, which led to the damage of the asphalt concrete on the bridge deck (about twice a year for maintenance). The above problems can be solved by changing the structural system or the welded steel plate on the bottom plate of the steel beam into a steel box girder, increasing the longitudinal bending rigidity and reducing the vertical deformation of the bridge. Therefore, to think about the road maintenance from another perspective, there must be new findings, making it necessary for cadres and workers of the highway system to re-study the rational use of limited funds.

(In the picture, it is shown that the collapse accident was caused in the demolition project of the cave, which should bring the attention to safety issues in the maintenance and even demolition of the arch bridge, especially the double curvature arch bridge.)

**Fig. 5.69** Photo of the overall collapse during the demolition of a cave in a province

(In the picture, it is shown that the whole arch bridge collapses into the water damage caused by the mountain torrents damaging the soil embankment.)

**Fig. 5.70** Photo of the overall collapse of an arch bridge into the water damage along a provincial highway

## 5.6 Similarity of Foundation Pit or Foundation Collapse

(1) Ground subsidence occurs nearby during the construction of an underground garage (conditional temporary system of equilibrium) (Fig. 5.75).

(According to the results of the preliminary investigation released by the government, the accident exposed potential production safety hazards in some areas: the troubleshooting and governance are not complete, and there are still weak links in the management of vehicle overload and overrun.)

**Fig. 5.71** Photo of the overall collapse of a provincial highway bridge under overload conditions

(In the picture, it is shown that the overload of coal transportation vehicles results in the fracture of simply-supported beams. Therefore, the design principles and requirements, especially the load standard, must be taken into account in the operation. It must be strictly prohibited for illegal vehicles to pass over the bridge.)

**Fig. 5.72** Photo of the partial collapse of a provincial highway bridge under overload conditions

(The bridge has been demolished and rebuilt to avoid possible accidents due to failure.)

**Fig. 5.73** Removal and reconstruction of similar highway bridges in a province to avoid possible accidents due to failure

**Fig. 5.74** Photos of pavement maintenance from the perspective of foundation deformation

Two months after the construction of the underground garage, the residents found that there were many cracks under the building, and the construction party filled them with cement many times. There were many cracks of different sizes in the left side of the residential building in contact with the construction site. The widest part was about 2 m, and the depth was about 3 m. As a result, the buried water pipe burst, and the whole area involved in the collapse was more than 10 $m^2$. Fortunately, there were no casualties, otherwise it would not be so simple. "The two examples of construction accidents are too similar."

**Fig. 5.75** Ground subsidence caused by underground garage construction

(2) A large amount of waste water constantly scours the foundation of the Sandburg, resulting in the destruction of several buildings (conditional temporary equilibrium system).
The Jesemel sand castle in India (Fig. 5.76) is a desert castle with a history of more than 800 years. Due to the development of tourism, many hotels and entertainment places have been built in the castle, and residents have begun to use convenient tap water. However, due to the lack of perfect drainage facilities, a large amount of waste water flows through simple drainage ditches and constantly scours the foundation of the sand castle, resulting in the collapse of

**Fig. 5.76** An 800-year-old desert castle in India

some sand and stone walls. Originally, 99 fortresses were preserved on the city wall, but now several have been destroyed.

(3) Collapse of nearby buildings during the construction of an underground garage (conditional temporary system of equilibrium) (Figs. 5.77, 5.78 and 5.79).

(In the picture, it is shown that at around 6:00 on June 27, 2009, a 13-story commodity building under construction collapsed, causing the death of a worker.)

**Fig. 5.77** Collapse of nearby buildings caused by underground garage construction in a city

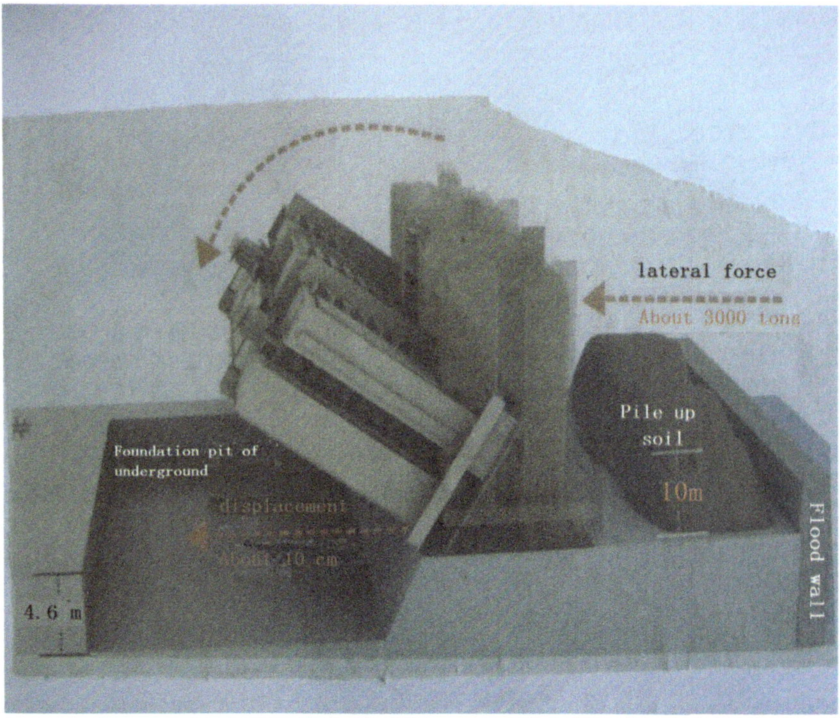


(In the picture, it is shown that the main cause of the collapse of the house was the instability of the side wall of the underground garage under construction and the secondary cause of the collapse of the house was the low resistance of the lateral force and PHC pipe string to the side; just like the demolition by blasting, the reverse direction can be determined by blasting the bottom of the collapsed side of the house first.)

**Fig. 5.78** Diagram of the analysis of the collapse of nearby buildings caused by an underground garage construction in a city

According to the results of the survey released by the government, the main reason for the house dumping is that the soil piled on the north side of building 7 became too high in a short period of time, with the highest point of about 10 m. At the same time, the foundation pit of the underground garage close to the south side of the building was being excavated to a depth of 4.6 m. The difference in pressure on the two sides of the building caused the horizontal displacement of the soil mass, and the excessive horizontal force exceeded the side resistance of the pile foundation, resulting in the collapse of the house; the original survey report met the requirements of the specification after supplementary survey and review on site; the original structural design met the requirements of the specification after review. The PHC string used in the building met the requirements of the specification after inspection.

(In the picture, it is shown that the support of the underground foundation pit in a city is stable, which ensures the stability and safety of nearby buildings and is also the key to the problem.)

**Fig. 5.79** Underground foundation pit under construction in a city to ensure the stability and safety of nearby buildings

**Fig. 5.80** Photos of thc Shenzhen Futian underground railway station under construction

For example, the Shenzhen Futian underground railway station under construction (Fig. 5.80) is a large-scale comprehensive passenger transportation hub of a three-layer structure connecting the Pearl River Delta region. There are many technical problems, such as making full preparations, scientific planning and design, risk assessment, scientific construction design plan and safety accident prevention plan, etc., which create a situation of progress without serious problems.

# Chapter 6
# Suggestions for Typical Problems

## 6.1 Monitoring of the Stability of the Structural Branch Point

In the process of construction and use of engineering structures, there are often states of metastable equilibrium (state 2) from a state of stable equilibrium to an unstable equilibrium one. As shown in Fig. 6.1, the pile foundation of structure (a) and structure (b) are all in the foundation pit. Structure (a) shows that the crown beam is connected with the column by a hinged connection, and structure (b) shows that the crown beam is connected with the column by compression only without tension support. When there is external force on this kind of structure, structure (a), the ultimate failure will occur only after the external force reaches the ultimate value, and the whole structure will not change in this process, so it belongs to the problem of extreme point instability; on the contrary, when subjected to external force, the columns in structure (b) in the foundation pit have the deformation trend as shown by the arrow, so the deformation often occurs in the whole structure before the structure reaches the ultimate bearing value, This kind of problem belongs to the problem of branch point instability.

The stability of branch points is often ignored in the design and construction of engineering structures, as shown in Fig. 6.2. For example, some accumulated damage to bridges may change the stability of the extreme point of bridge stability into the stability of a branch point; for example, some underground projects seem to be the stability of the extreme point from the perspective of stratum structures, but some supporting structures have characteristics of obvious brittle failure, which can cause an avalanche type chain failure, which belongs to the stability of an obvious branch point. In view of this kind of problem, it is difficult to control the overall stability of the supporting structure of the stratum even if a monitoring measurement is used, so close attention must be paid to it. Only by improving the supporting structure measures and transforming the stability of the branch points into the stability of extreme points, can the monitoring and measurement be effectively carried out and the overall stability



H. Zhu and L. Shi, *Methodology of Highway Engineering Structural Design and Construction*, Advanced Topics in Science and Technology in China 59,
https://doi.org/10.1007/978-981-15-6544-1_6

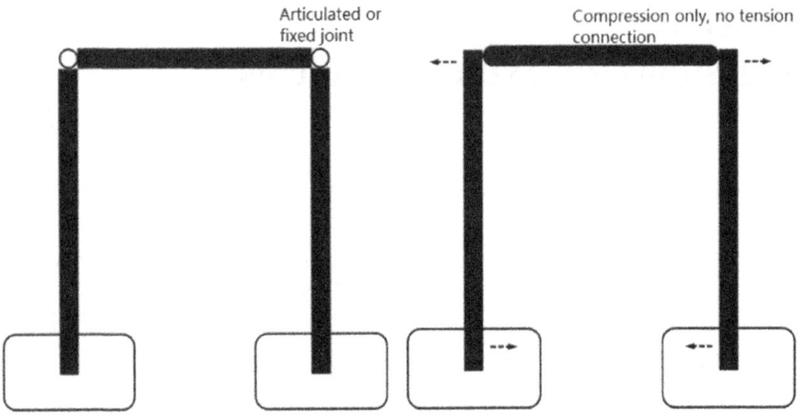


a) Structure of the extreme point instability b) Structure of the branch point instability

**Fig. 6.1** Diagram of the stability type of the engineering structure in a state of equilibrium

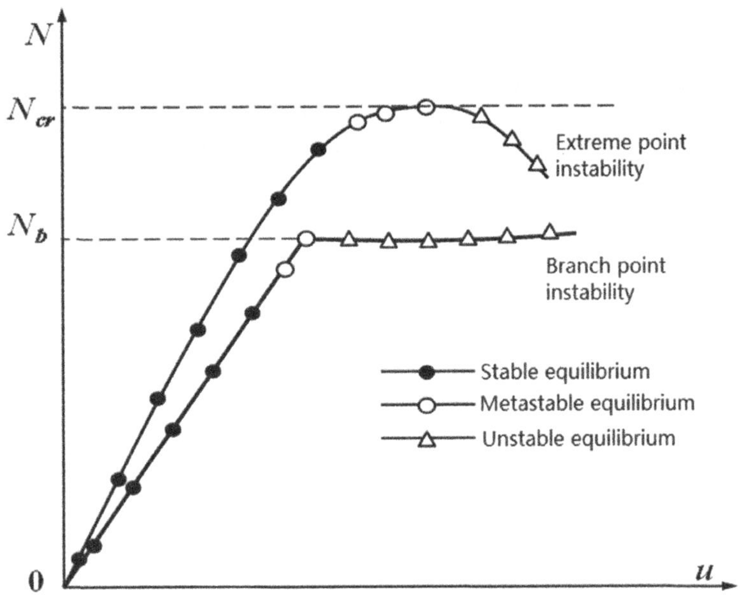


**Fig. 6.2** N–δ curve of the stability of engineering structures in a state of equilibrium

of the supporting structure of the stratum be controlled. **In particular, with the excavation process, the supporting structures such as large-span tunnels with unfavorable geological surrounding rock, deep and large foundation pits in soft soil, etc. are in an uncertain state, which requires a reasonable system of supporting structures to ensure that the state of the system of overall (stratum**

**Fig. 6.3** Site of the collapse of a deep foundation pit and damage to the supporting system

**support) structure remains basically unchanged.** ***It can be seen that monitoring is only applicable to the stability of extreme points, but not to the instability of branch points***. In practical engineering, when we plan the scheme of the engineering design and construction, we should adopt reasonable structural measures or auxiliary means to prevent structural instability and to ensure the safety of the engineering construction.

For example, a deep foundation pit witnessed a serious accident of collapse (Fig. 6.3). The key point is that the condition of equilibrium and stability of the structure of the foundation pit had adverse changes during the process of construction. As shown in Fig. 6.4, it can be seen from the supporting system before the foundation pit was damaged that if the two sides of the foundation pit do not rotate, the supporting system will effectively provide the supporting force. However, this can be clearly seen from the rotary deformation of the side wall of the foundation pit after the failure (Fig. 6.5). Due to the rotary deformation displacement of the side wall of the foundation pit during the construction and excavation, the condition of equilibrium of the foundation pit supporting system was damaged (**on the surface, the whole seems like the instability of the extreme point, but in fact, there was the instability of the branch point in the key supporting structure (supporting steel pipe)**).

In case of damage to the foundation pit, the method of benching tunneling is adopted for excavation, and the bottom of the pit is sealed section by section, as shown in Fig. 6.6. With the advancement of the process of excavation, the scope of the bottom of the foundation pit is expanded. Due to the poor mechanical properties of the soil in the foundation pit, as shown in Fig. 6.7, the soil deformation at the bottom of the deep foundation pit changes the direction of the force transmitted downward by the original structural load, forming a downward and lateral coexisting force. However, the supporting system of the deep foundation pit does not have integrity, and cannot control the impact of the unstable mud at the bottom of the foundation pit, resulting in the rotational deformation of the side wall of the foundation pit (Fig. 6.7).

**Fig. 6.4** Part of the supporting system before failure

The upper supporting rod changes from the stable state of a compression load to the state of tension and separation, which causes the failure of the supporting system.

In the process of the design and construction of a deep foundation pit in a poor and weak geological environment, it is very important to control the instability and mud inrush at the bottom and the integrity of the supporting system. The core of the serious accident due to a collapse of the deep foundation pit was that the combination system of rock and soil mass and the supporting structure lacked sufficient ability to control the deformation coordination. This project belongs to "in the actual construction process of foundation pit, there is a metastable equilibrium state from stable equilibrium to unstable equilibrium". In fact, the monitoring of point A on the surface of a foundation pit was not reliable, and it was difficult to monitor point B in the process of excavation. Only when the engineering design and construction scheme is studied, can reasonable structural measures or auxiliary means be used to prevent the structure from losing stability, and can the effective monitoring be carried out. Experts put forward three principles that must be followed in the construction of the deep foundation pit: ① The excavation of the foundation pit must be layered and segmented, and the excavation exposure time should not be too long, each layered excavation should be controlled at 3 m, and the segmented excavation should be guaranteed at 15–20 m; ② The foundation pit must be supported before excavation, and the supporting details should be well grasped, and the deformation of the foundation pit should be controlled; ③ Pay attention to the foundation on rainy

Fig. 6.5 Rotational deformation of the foundation pit side wall

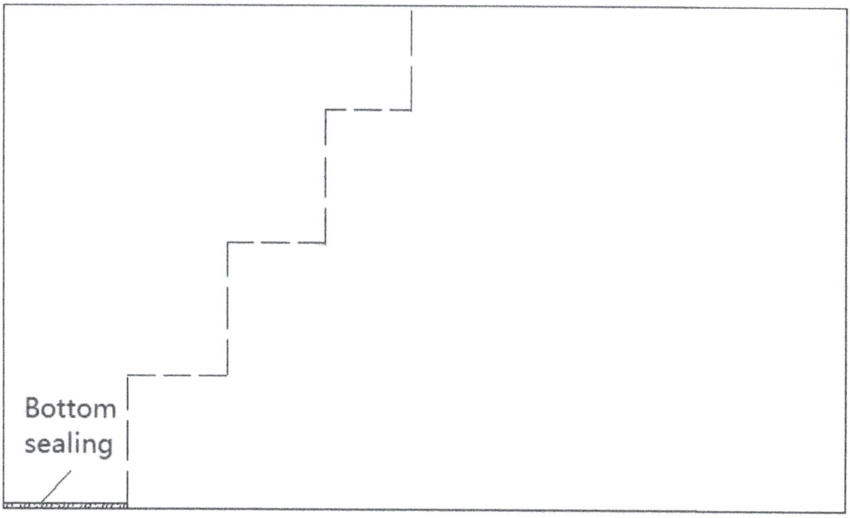


Fig. 6.6 Excavation by the method of benching tunneling, section-by-section bottom sealing

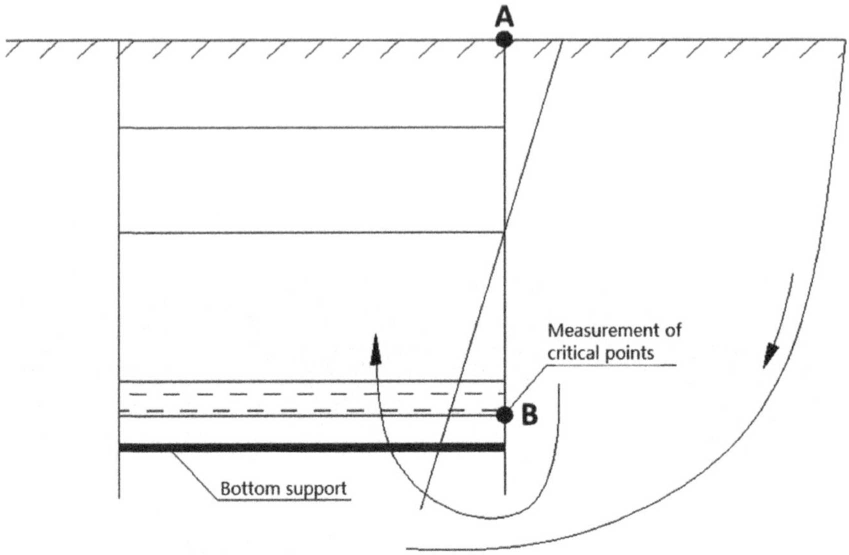


**Fig. 6.7** Diagram of the analysis of the mechanism of the instability of a deep foundation pit

days and the pit should be drained in time. After completion, the concrete should be reinforced immediately to ensure that the pit does not deform. These three principles belong to the empirical category, without indicating the concept of structural mechanics, and basically meet the requirements of the stability of the foundation pit and the supporting system in a state of equilibrium. There are risks in the operational process.

## 6.2 Potential Problems in the Utilization of the Plastic Hinge of Cyclic Load Structures

The national statistical results show that: The collapse of roads and bridges accounts for 9.9%, but the number of critical bridges whose service life is less than 30 years accounts for 64% of the total number of critical bridges; overload is only an inducement, **the key is that the structural design is weak or unreasonable, which aggravates the fatigue damage or lateral displacement of the structure and shortens the service life.** This is consistent with the statistical results that the types of bridges in recent years are classified as simply-supported slab girders, composite arch bridges, rigid frame bridges, etc., all of which belong to a flexible structure or a rigid flexible composite structure. Because this kind of structure is often in an elastic-plastic working state, but also subjected to high cyclic stress or unreasonable stress, the process of stress does not have superposition, the state of stress of the structure is related to the loading path (in a metastable state of equilibrium), just the variability of the repeated load path of the automobile is not as stable as the repeated vertical load path of the fire truck or industrial crane. In this way, the design stress of the flexible structure and the rigid flexible composite structure may differ greatly from the actual stress under the repeated load or even partial overload of the vehicles, and it is easier for the structure in the elastic-plastic working state to produce more accumulated damage than the structure in the elastic state, and the adverse forces and energy are transferred or concentrated on the elastic-plastic weak part of the structure, so as to reduce the service life of the bridge, which may lead to bridge damage or periodic damage in advance, and there is a risk to the safety of driving across the bridge. However, according to the repeated design of load bridges of trains or industrial cranes, under the action of an elastic stage, the structure is in the working state of elastic stage, the design stress is basically the same as the actual stress, and the risk to the safety of driving is small (in a stable state of equilibrium); for example, the weight of a 20 m standard plate beam is 26 t, while the maximum specified weight of the vehicle is 55 t, and the bearing capacity of each plate can be simplified to about 27.5 t (1/2). Especially under the condition of overloaded operations, the structure is often in the elastic-plastic working state, the cumulative damage to the structure is large, and the risk to the safety of driving is large (in a metastable state of equilibrium). The Qiantang River Bridge designed and constructed by Mao Yisheng, a famous bridge engineer, is a classic and mature steel truss bridge. It is controlled by structural rigidity, with surplus safety, stable quality of the material, basically in the elastic working state, less cumulative damage to the structure, and less risk to the safety of driving (in a stable state of equilibrium).

In the past, the non-elastic parts of small and medium span bridges and some special components of bridges would accelerate the accumulation of damage under repeated overweight loads, and especially in the 1980s, most of the optimized bridges had problems. It was difficult for the previous method of design to avoid such problems, which is also one of the reasons that overloading easily leads to the collapse of highway bridges. The core of the problem is that it is difficult to guarantee the adaptability of the structural transmission medium and avoid the metastable equilibrium of the structure.

Many small and medium-span bridges are prone to damage due to the accumulation of fatigue. It is difficult to explain why many small and medium–span bridges will collapse suddenly under the action of heavy traffic vehicles by static analysis alone. It is necessary to combine the theory of damage due to the accumulation of fatigue and the "principle of minimum energy consumption" for an analysis. In fact, only the ideal elastic body can satisfy the relationship that the external work equals the elastic strain energy of the structure, while the non-elastic body can generate the accumulated damage energy of the structure in addition to the elastic strain energy to cause material degradation (Fig. 6.8).

The structure designed according to the construction traffic code is considered to work in an elastic-plastic state. For long-span highway bridges, the weight of the bridge itself is the main part, and the live load is relatively small. Under normal use, the structure tends to work in an elastic state; for medium and small-span highway bridges, the weight of the bridge itself is equivalent to the live load, and under normal

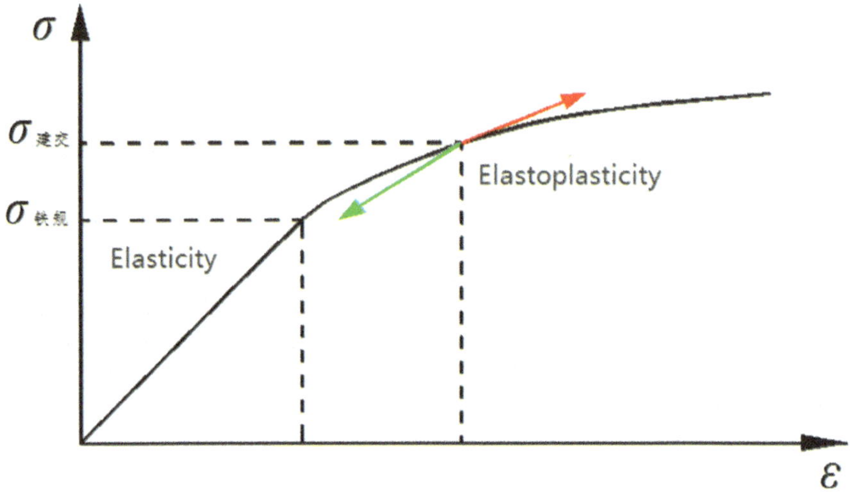


**Fig. 6.8** Comparison of the damage due to cumulative non-elastic volume under static load and repeated load

use, the structure tends to work in an elastic-plastic state. In fact, the main highway bridges often bear heavy traffic or even part of an overload, which can easily produce cumulative damage under repeated loads; while the structures designed according to the railway code only allow the structure to be in an elastic working state, and the railway load generally does not appear in overload, so the cumulative damage under repeated loads is relatively little (Fig. 6.8). In addition, when calculating the bending capacity of steel beams that do not directly bear the dynamic load according to the building code (later changed to "need to calculate fatigue"), it can be considered that the section part enters into the plastic state, i.e. according to the elastic-plastic theory, but when directly bearing the dynamic load, the code can only design the bending capacity according to the elastic theory. Correspondingly, the international organization for Standardization (ISO) has made two regulations: ① Plastic design cannot be used for members with alternating plasticity, i.e. no yield in tension or compression; ② For the structure under dynamic load, the design load must not exceed the stable load, that is, the component will not be damaged due to the gradual accumulation of plastic deformation, nor will the material produce low cycle fatigue failure due to alternative tensile yield and compression yield. There are occasional accidents involving a collapse of main highway bridges around the world, not railway bridges. In addition, only a few types of bridges in the same section of a highway collapse after a period of operation, and all bridges are in a similar state of partial overload. Just as it takes only a few back and forth actions to break the wire in life, the reason is that each back and forth action has reached the elastic-plastic state, and the adverse force and energy are transferred or concentrated to the weak elastic-plastic part of the wire. The above two situations show the importance of the main highway bridge structure in an elastic working state.

The core of the application of the method for the control of deformation coordination in the design of highway bridges is to ensure that the structural forces, such as bridges, are transferred according to the design path, and that the control forces are unreasonable or even harmful. According to the coordinated control method of structural deformation, the design improved, the rigidity of medium and small-span bridges moderately improved, and the compatibility of the deformation of some special components of bridges improved. Under the repeated overweight load (Fig. 6.9), the accumulated damage of the non-elastic parts will be in a controllable range, avoiding the problem in which the appropriate overload will easily lead to the collapse of highway bridges. It can ensure the adaptability of the medium of structural transmission, avoid the problem of structural metastable equilibrium, and ensure the safety of the engineering structure.

The plastic hinge of structural mechanics can be effectively used in metal processing, but the use of it in bridges will lead to the problem of accumulated damage, which should be considered!

For example, the destructive test results of a 20 m hollow slab beam structure are shown in Fig. 6.10.

It can be seen from the destructive test results of the optimized wide-width hollow slab beam structure that the small hinge joint can easily be damaged (secondary cause) due to the small amount of stiffness (main cause), just like the innate immune capacity of the human being is poor, one easily gets sick or acquires small diseases soon after; especially when the main highway bridge often bears heavy traffic or even is under partial overload, the structure changes from an elastic working state to an elastic-plastic working state, which can easily produce accumulated damage.

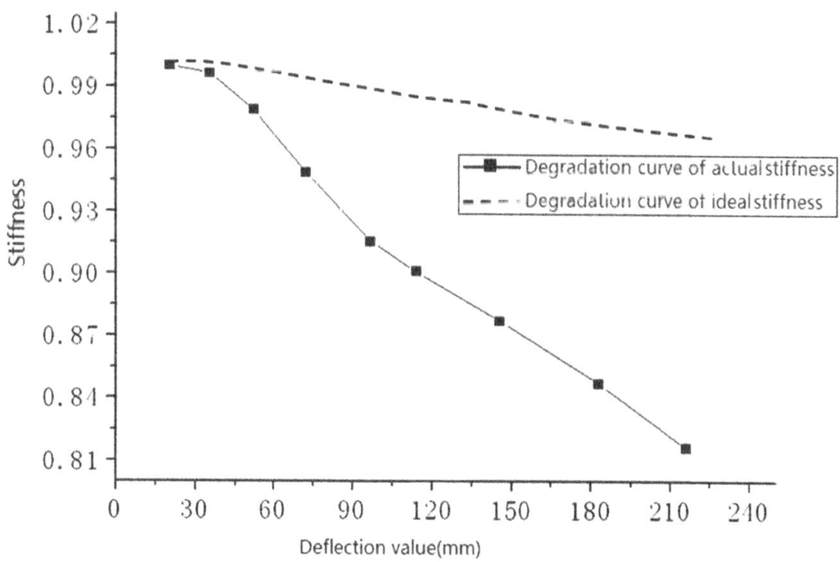


**Fig. 6.9** Stiffness change during loading

**Fig. 6.10** Photo of the collapse of a bridge with 20 m hollow slab beam

Therefore, the vertical stiffness of a 20 m simply-supported hollow slab girder is relatively small, which is only suitable for the vehicle action within the standard load of a volume of the general traffic of an ordinary highway; it is not suitable for the overload action of a large volume of traffic or even part of the overload action of main highway bridges. This is consistent with the case in which the simply-supported hollow slab bridge beam of 20 m on the main road often bears heavy traffic or is even under a partial overload, which can easily be damaged or broken or even affects the safe operations.

# Chapter 7
# Guiding Documents

## 7.1 The Philosophy Behind the Building Up of Water Conservation

**Pan Jiazheng**

**Editor's notes**: The building up of water conservation is an important aspect of the construction of national economic infrastructures. To master a great deal of modern scientific and technological knowledge, water conservation engineers need to think comprehensively and from a higher level. Based on more than 50 years of experience in water conservation and hydro power construction, the author analyzes and considers several aspects of the build-up of water conservation from the perspective of philosophy. With the academic thoughts of ancient and modern Chinese and foreign sages and the practical examples of daily life, this paper makes a historical, dialectical and materialistic reflection and elaboration on the advantages and disadvantages, success and failure, cost and benefits, certainty and risk, refinement and synthesis of methods, the strength and weakness of materials, the coarseness and fineness of norms, avoidance and advantages and disadvantages, prudence and innovation of the construction of water conservation structures.

There are many subjects for hydraulic engineers to master, so it is impossible to spend a lot of energy on studying philosophy. But one's thoughts, words and deeds are always dominated by one's own epistemology and world outlook. If there is a deviation in these aspects, although you have a good intention to master modern knowledge of science and technology, it is often half the effort with half the effort, and even leads to unexpected consequences. In this way, it is beneficial for water conservation engineers to read some philosophy books. However, engineers do not have to read classics. Sometimes, a high-level philosophy paper does not work as well as a proverb. The eight questions mentioned below are some of my sporadic feelings in my career of more than 50 years of water conservation and construction of



H. Zhu and L. Shi, *Methodology of Highway Engineering Structural Design and Construction*, Advanced Topics in Science and Technology in China 59,
https://doi.org/10.1007/978-981-15-6544-1_7

hydro power centers, and these questions are only very superficial. They are written for your reference. Please correct the defects.

**The Philosophy of Looking in the Mirror**

That is to say, in the *A Dream of Red Mansions*, a mirror should look at both sides. Taking water conservation as an example, human water control has a long history. Especially in China, floods and droughts are especially serious and frequent. Since the recorded story of "King Yu Tamed the Flood", the history of the development of Chinese civilization is almost a history of water control. There are both successful experiences and lessons from failure.

In the early days, people had no choice but to escape from the vast floods or long-term droughts. If you cannot avoid it, you will become a corpse of starvation. Since then, with the development of science and technology and productivity, and the accumulation of experience, people have begun to subdue water, open channels, dig wells, repair embankments, and build dams. The scale and role of the project has been expanding, reaching a climax in the last century. In this contest, mankind seems to have won a great victory. For example, China has made remarkable achievements in flood control, irrigation, water supply, power generation and other aspects in the past 50 years, but there have also been a series of mistakes, which must be observed at the same time.

Specifically, in the past 50 years, we have built 260,000 km of river levees to ensure the safety of the Yellow River and the Yangtze River; we have built 85,000 reservoirs, drilled millions of machine wells, and increased the national water supply from 100 billion cubic meters in 1949 to 560 billion cubic meters every year today, ensuring the development of industry and agriculture. The irrigated area has grown to 53.74 million hectares, with limited cultivated land supporting more than 1.2 billion people, and the output value of the GDP is over US$1 trillion; more than eight million kilowatts of hydro-power have been developed and a Grand West to East power transmission project is under way. The Three Gorges water conservation project under construction is the largest water conservation hub in the world and will start to play its benefits this year. More projects, including the famous "South to North Water Diversion Project", are under construction, in preparation or in the planning stage. The height of the dam has reached 300 meters, which seems to have achieved the level of "everything becomes possible".

However, in the shadow of victory, sober people see some thought-provoking problems. Large reservoirs and long causeways have been built to control the floods, but the causeways are getting longer and longer, the riverbed is silting up, the flow capacity is shrinking, and the situation of low floods and high water levels appears. In 1998, the flood discharge of the Yellow River at Huayuankou was only 7,600 $m^3/s$, and the water level was 0.91 m higher than 22,300 $m^3/s$ in 1958. The same is true for the Yangtze River. During the flood season, there are many dangerous situations, and tens of thousands or even hundreds of thousands of people go to the levee to be rescued. Is there an end to such an "ever higher embankment for flood-fighting"?

The water supply is increasing exponentially and the economy is developing greatly, but the cost is: excessive development of water resources, inefficient utilization and serious waste, river channel drying up and groundwater level dropping substantially. If China wants to catch up with the developed countries, its GDP will have to increase more than 10 times. Where can we get the water we need?

The cascade development and channelization of rivers have brought great benefits, but for some rivers, natural floods have disappeared, fish migration channels have been cut off, some species have become extinct, and some reservoirs have gotten silted up.

The most serious problem is water pollution. Some people describe it as "All water is dirty, all rivers are dry". It is said that our development is to "Eat the food of our ancestors and cut off the food for our descendants". Now there are some organizations and people around the world who are against the construction of all dams and water conservation projects. We firmly oppose radical statements and practices that ignore national conditions and make up "stop eating because of choking". However, it is absolutely necessary to conscientiously sum up experiences and lessons, improve work and avoid mistakes.

I think the correct attitude is to remember Lao Tzu's saying: "Happiness breeds misfortune". There are always two sides to everything in the world. After millions of years of evolution, nature maintains a relatively balanced form. The construction of water conservation projects, especially large-scale projects, will inevitably break the balance, cause a series of disturbances, and reach a new balance after several years. In the process of disturbance, there are always gains and losses. There is no perfect good in the world. The problem is to judge the merits and demerits scientifically and fairly, and everything must be measured from a long-term and overall standpoint. The loss should be eliminated or compensated for as much as possible. This is the responsibility of hydraulic engineers.

We must oppose departmentalism and short-term views. Nowadays, in order to build a project, we have to do a "feasibility study", but generally speaking, the benefits of the project are always discussed repeatedly, and the side effects are always ignored. It is inevitable that officials should reflect their achievements; business departments should develop themselves; design companies and construction enterprises should solicit living things to eat. It is also too harsh for them to give up their local, standard and recent views completely. Only God considers the whole situation. This God is the state, the government and the detached scientific community.

Special attention should be paid to the following issues:

When building a flood control reservoir to control floods, the downstream area cannot occupy the flood discharge space without a limit. To realize that, people cannot eliminate floods, we must learn to coexist with them and leave the necessary flood discharge space to floods.

In order to solve the demand for water of industry, agriculture and urban life, we should not open up the supply "according to the demand", and we should not supply water at a low price or free of charge. It is a big principle to keep the equilibrium between pumping and storing during a long period of time. In areas where there is a water shortage, in order to maintain a certain pressure of shortage is to implement high price water supplies and not to "benefit the people". Otherwise, a big water supply means a big waste, great damage and a great deal of pollution. Today, water resources in some areas have been seriously damaged, the social habits of water waste cannot be reversed, and water conservation engineers have inadvertently played a role in this.

Don't think that adjusting the runoff of a natural river channel to a uniform lowering, eating and drinking is the best plan. We should carefully study all of the side effects, release artificial floods when necessary, scour riverbeds, form waterfall rapids, and make sure that a certain amount of water returns to the sea.

The construction of water conservation projects should be coordinated with the departments of ecology, the environment, navigation, tourism, cultural relics, etc., with special attention to pollution. In principle, where water pollution has not been treated and water waste has not been solved, development projects should not be carried out. In the arid inland areas, it is not permitted to store water, reclaim wasteland, plant trees and make artificial oases. For the northwest, we can only adapt to the natural conditions, guide ourselves according to the circumstances, protect and improve the environment, and do not imagine transforming it into a thousand miles south of the Great Wall. "The sky is gray, the wild is boundless, the wind blows grass to see cattle and sheep" is also the scenery that must be preserved.

Let's remember the wise saying that "happiness breeds misfortune", and of course, we should also remember the saying that "misfortune generates happiness". As long as we discover the mistakes of the past and improve our thinking, actions and policies, water conservation projects can really achieve the goal of promoting the benefits, eliminating the disadvantages, avoiding the disadvantages and benefiting the people. In the new century, they will be carried out on a larger scale and will reach new achievements.

**The Philosophy of Taking a Flight**

If there are two planes, one of which is very new, one of which is out of service, and there was an accident not long ago, which one would you choose? Believe that it is easy for people to make a choice. Even if the owner of the old plane has invited an expert to make a diagnosis and assessment, and thinks that the plane is still in line with the flight conditions, I'm afraid that it would not change your decision, because the risk of taking the old plane is certainly greater. Of course, if the mission is urgent and there is no other means, we can only take the old plane.

There are also problems in the understanding of "risk theory" or "determinism" in water conservation projects.

Many young engineers see work as a matter of certainty. They have modern scientific and technological knowledge and means, can carry out complex analysis and calculation tests, and are familiar with rules, specifications and standards. But they ignore one point: Many basic data, parameters, assumptions, methods, etc. are quite arbitrary. The normative provisions are only a summary of past experience, and there is no absolute correctness. For example, when designing a dam, we cannot find all the conditions in the foundation, fully grasp the characteristics and response of the materials, and we cannot predict what kind of flood will be encountered and how strong an earthquake will be induced after completion. For example, we can now calculate the response of an arch dam during an earthquake very accurately, even the opening and closing of the transverse joints can be considered, but all of the results depend on the given ground motion process, which is unknown. We can only assume many lines of the process in order to do the research. In hydraulic design, to a certain extent, we have to rely on past experience and the judgment of engineers. Therefore, the design based on modern scientific theory, analysis and tests by modern means, and meeting the requirements of regulations and norms can stand on the legal principle, but no designer has the right to declare that his design has no risk.

I deeply feel that in the design of water conservation projects, "determinism" has a greater impact, and this may be related to school education. The teacher always asks the students to come up with a definite answer. Some leaders often ask experts to come to simple conclusions, for example, "Can high dams be built on this dam site?". Experts also dare to make certain answers: "xx dams of a height of xx can be built". I always have doubts about such questions and answers. With the development of science and technology and the enhancement of economic strength, regardless of the risks and the investment made, it seems difficult to exclude the possibility of building a dam on a certain dam site. On the contrary, it is too easy to build a dam without additional constraints.

In fact, there is no design or engineering without risk, but the risk is high or low, and the consequences of engineering failure are only large or small, which requires decision-makers to make a comprehensive measurement. Still take flying as an example. If you have to arrive at a place in the shortest time, other means of transportation cannot satisfy you. There is only one plane in the airport, and the plane has been verified, then taking this plane is a reasonable decision. However, if there is no such urgent need, there are many means of transportation or many planes to choose from, and you are carrying a family of men and women on board, then it is worth pondering whether you have to take an overaged plane that has something wrong.

Take the dam site selection as an example. If we want to build a beneficial reservoir, there are two dam sites to consider. The geological condition of dam site A is poor, there are active faults nearby, the seismic intensity is high, but the reservoir capacity can be larger and more benefits can be obtained. The geological condition of dam site B is good, but it is located in the upstream area and the storage capacity is small. Which plan should be selected? The essence of this problem is to balance and choose among benefits, costs and risks. The answer depends on the specific conditions and there is no rule to follow. We must make several basic problems clear. For example, in the aspect of benefit comparison, whether plan B does not meet the basic requirements of database building, or is only "relatively poor". In terms of risk comparison, what is the probability of a dam breaking and the consequences of a dam break in plan A when it encounters a strong earthquake. If the probability of the dam breaking is very low and the consequences are not serious (no death or few casualties), the problem of risk for the project does not need to be placed in an important position in the decision-making process; on the contrary, if the dam breaking would cause the destruction of the downstream towns and heavy casualties among the people, the problem of risk would become a major factor to consider in making the decision. In this comparison, there is no absolute standard, but a trade-off between benefits and risks, and a technical and economic comparison is not the decisive factor. Because the assessment of the risks can only be a fuzzy concept, it is often difficult to reach a consensus on such issues, and the final choice often depends on the manner of thinking and a coordination of the results at the decision-making level.

**The Philosophy of Taking Traditional Chinese Medicine**

When I was young, I relied on traditional Chinese medicine. When I grew up, I changed to Western medicine and took Western medicine. In the impression, there are many superstitious and backward elements in traditional Chinese medicine, and many theories are not scientific enough, while Western medicine is much more scientific and credible. The development of Western medicine is more rigorous. A kind of medicine has a specific effect and function. Its molecular structure is clear. How the body absorbs and excretes it is also clear. This is not comparable to the traditional Chinese medicine that boils the bark and grass roots in one pot. However, now my understanding has changed. Although Western medicine is scientific, it seems to lack the spirit of "synthesis" and "dialectics". Although traditional Chinese medicine is "unscientific", many of its principles, such as "dialectical diagnosis and treatment", "comprehensive care", "matching between the monarch and the minister", "adjustment according to specific conditions" and so on, are really meaningful. If we combine the advantages of the two, we will reach a new level. In water conservation projects, there will also be the situation of "taking traditional Chinese medicine".

For example, in the design of the grade of concrete for a dam, we divide the body of the dam into several areas, and put forward the requirements of strength, age, impermeability, abrasion resistance, durability, slump, water cement ratio and so on, which seems very scientific, but the results are as follows: Sometimes, the performance of concrete in adjacent areas is quite different; the "abrasion resistant concrete" is completely cracked; in order to resist seepage, the amount of cement should be very high, which is not suitable for other requirements. All of these are unreasonable and will have adverse consequences. This is the loss of "having blind faith in Western medicine". Many people desperately take measures to reduce cholesterol, but the result is that they have destroyed their health.

As an example of the optimization of an arch dam, there are two kinds of understanding. According to the route of Western medicine, first establish an objective function (for example, take the total volume or total cost of the arch dam), then determine some conditions (called constraints) that must be met in the design and construction of the arch dam, and then establish a mathematical model that can describe the shape of the arch dam, including the dry design variables, and changing these variables, you can get different shapes of the arch dam. At last, by using the method of mathematical programming, a set of design variables which can satisfy all constraints and make the objective function take the minimum value are found, and the optimal plan is found. Its idea and technical route are beyond reproach, but sometimes the optimized cross-section is not reasonable, and the results are different depending on different assumptions, and there is no absolute optimal solution. Second, whether the objective function should consider anything besides the pursuit of the minimum engineering quantity will lead to the problem of multi-objective optimization. For example, make the quantity V and engineering safety K reach the most appropriate state at the same time. Generally speaking, with the increase of V, K also increases. The relationship between V and K can be drawn into a curve. It is a feasible region between K greater than the minimum safety factor and V less than the maximum allowable quantity of work. The problem lies in how to integrate the factors with different properties into one to select an optimal solution. After all, economy and security are not of the same nature. Furthermore, it is difficult to find a reasonable and recognized evaluation index to measure the so-called engineering safety K. It seems that a problem of "fuzzy comprehensive evaluation decision-making" cannot be solved by simple mathematical optimization. Maybe a meeting should be held based on a great deal of analysis and calculation results in order to make a decision, which in some way is like decoding Chinese medicinal herbs.

In demonstrating the feasibility of the Three Gorges project, a famous Canadian hydro-power consulting company conducted an independent demonstration under the guidance of the World Bank. I have been working with Canadian experts for several years, and I feel that they are highly skilled, experienced, efficient, and scientific and I have learned a lot from them. But they also feel that they pay too much attention to the analysis of economic benefits when they choose or decide, which is a typical school of "Western Medicine". Of course, we should pay attention to the input-output problem in the construction of water conservation projects, but the project benefits have both specific economic benefits and social benefits. How can benefits such as "reducing the possibility of flooding cities and people ", "relieving people's psychological pressure" and "protecting rare species" be converted into dollars? Canadian experts advocate reducing the flood control capacity of the Three Gorges reservoir, allowing the people in the reservoir to escape temporarily and making compensation afterwards in case of a major flood, which can reduce the pressure of migration, and it is also economically beneficial. In fact, we initially envisaged this plan, but after an in-depth demonstration, many more difficult problems emerged that needed to be solved (such as whether the flood escape area can be constructed, how to develop it, whether the temporary flood escape will cause the death of people…). In the end, it was not suitable to adopt the plan according to the national conditions and public opinion. In the exchange, it is difficult for Canadian experts to understand and accept our views. Finally, I can only say: "Your analysis is very scientific and accurate, but we have to consider more factors when making decisions. We need to coordinate comprehensively. For example, when using traditional Chinese medicine to treat diseases, we need to increase or decrease the taste and dosage of drugs based on patients' conditions." After listening to this, the leader of the Canadian experts smiled bitterly and said, "I have been working in technology all my life, but you advised me to choose a traditional Chinese style of working!" Maybe this is the subtle difference between Eastern and Western cultures.

## The Philosophy of Holding Eggs

An egg is a fragile thing, not only "hitting the stone with the egg" means inevitable failure, as long as it is lightly hit on the desktop, the egg shell will break and the egg yolk will flow. However, it is not easy to crush an egg if you hold it in your hand. If the egg is buried deep in the sand, a truck could even drive over it and the egg would still be safe. God is a great designer who can use such a little material and a beautiful and reasonable body shape to make the eggshell have such a strong resistance.

In fact, not only eggs, but also other soil, sand, rocks, slopes, etc. As long as they are good at understanding their nature and using their strengths, they will play their potential beyond imagination. Moreover, with the deterioration of the situation, it will continue to adjust itself and tap its potential. It will never fail until the potential is completely exhausted. It is a pity that we often fail to realize this, fail to use our strengths and make something die prematurely. It is like not holding an egg in your hand, but putting it on a stone, and blaming the eggshell for being too thin.

Sometimes, people not only do not take advantage of the material, but they also regard it as an enemy. In the early years, when we designed underground structures, we focused on lining and reinforcement. For the surrounding rock, it was mainly a question of calculating the pressure of the rock, its elastic resistance and so on. In fact, with a reasonable tunnel shape and method of construction, even the poor surrounding rock has a very strong ability to stabilize itself. If we help it (reinforce) according to the circumstances, it seems that the incomplete rock mass can form a thick natural arch, which is far stronger than the artificial lining arch. This truth is now more and more known. In order to adapt to nature, the flexible lining is more useful in replacing the rigid lining. Unfortunately, there are still some people in China who always discriminate against the natural surrounding rock and value the artificial lining. Even because of the unreasonable design and barbaric construction, the life of the surrounding rock is ruined, and they blame the poor geological conditions, which is unfair.

Let's study the principle of flexible support further. Hang some nets on the wall of the tunnel to spray some slurry, hit a few anchor bolts, and pour a little slurry. What's the effect on the earth? Some comrades made great efforts to do a theoretical analysis, and took the shotcrete layer and anchor bolt as structural elements for a simulation calculation. This is of course very useful, but I think there are two main functions of flexible support: one is to prevent local collapse, damage to the shape, and a continuation of deterioration, so as to lose the function of giving full play to the potential; the other is because the support is flexible, allowing the surrounding rock to have a little deformation, allowing it to adjust itself and tap its potential, these functions may not be calculated. In the past, there was a saying that "a straw killed a camel", which means that the load on the camel will gradually increase, and there will always be a limit from a quantitative change to a qualitative change. At that time, another straw could also crush the camel. Conversely, a straw can save a structure. The collapse of the cavern or slope always starts from the local small pieces falling or the local range staggering, and then it gradually gets worse. If you start with a little effort to prevent this from happening, the result is quite different.

A thin column is very flexible under pressure. If several fulcrums are added, the thin column can bear much more pressure even though it does not provide much reaction force. All of these are for the same reason, that is to say, it will take a very small cost to prevent the balance from inclining to one side and developing maliciously.

We can make the most economical, reasonable and safe design by fully mastering the material characteristics, guiding our plan according to the situation, tapping the potential, and living in harmony with it.

**The Philosophy of Eating Arsenic**

Arsenic is a highly toxic substance. If you take a little of it, you will be killed immediately, so that people shudder upon hearing the term. However, it has been recorded that some people are addicted to arsenic. They eat a small amount of arsenic every day and persist in it. They are not poisoned, and even live a long life. In their old age, they are still full of youthful vitality. For example, a centipede can be used as medicine and snake liquid can detoxify. Although these cannot be believed completely, there is no doubt that as long as we know the characteristics and control the dosage of highly toxic things, they are harmless or might even have good effects. In addition, the world's waste is countless, and to become the garbage everyone hates, the world spends a lot of money to deal with it. In fact, through skillful transformation, broken tree roots can become rare art treasures, and broken bricks and tiles may contain important cultural relics. In a word, the concepts of "poisonous" and "ragged" are relative. In many cases, there is the possibility of turning waste into treasure and winning with poison.

This is also true in engineering. In terms of "utilization of waste", the use of garbage to generate electricity is a clear proof of this. For example, the flyash discharged from coal-fired power plants has been treated as waste products for a long time, and there will be problems no matter whether it is piled up on site or discharged into rivers. It has been found that part of cement can be replaced when mixing concrete (or making cement), which not only saves cement (or clinker), but also has many advantages and can find a way for its large-scale utilization. At present, the high-quality flyash of some thermal power plants has become a hot product in short supply, and the price is also the same as that of cement. The saying that "All things in their essence are good for something." seems to be applicable not only to people, but also to other objects.

Take another example, that of MgO concrete. The cement or concrete contains a magnesium oxide component, which will produce volume expansion after setting and cause concrete cracking or even disintegration, which is an unstable factor. Therefore, it is regarded as a harmful component in relevant specifications, and its content is strictly limited. However, the effect of MgO on the expansion of the volume of concrete can offset or compensate the shrinkage caused by a temperature drop and the volume deformation of concrete after pouring. The latter is an important cause of the fracture of concrete buildings, especially structures of a large volume. Properly adding magnesium oxide to concrete can turn harm into benefit. The problem is to know its changing mechanism and control it strictly. The research in this field in our country is avant-garde. A large number of experiments and research projects have been carried out by the relevant researchers, the rules have been summarized, the measures have been put forward, and the practice has been carried out in the project successively, and good results have been achieved. There is such a project worth taking as an example: In an international bidding construction of a large hydro power station, due to various reasons, the progress of the construction was seriously lagging behind, which was considered as an incurable project by the World Bank. In order to recover the period of construction, the foundation concrete could only be poured on a large scale and at a high speed in the hot summer season, which is a great taboo in the engineering field. After repeated research, I decided to add magnesium oxide to the concrete to solve the persistent cracking, ignoring the strong written objection raised by the international group of experts, because I believe that the results of the research carried out by Chinese scientists who are willing to take responsibility for the grass-roots level and "dare to eat a little arsenic". As a result, the period of construction was miraculously saved and the dam has been running normally up to now. So let's not despise the rags and wastes around us. They are probably treasures.

**The Philosophy of Having a Physical Examination**

People cannot help getting sick. It is very important to have a check-up regularly. It is also understandable that important people and celebrities should be checked more frequently and carefully. Through physical examination, the situation of one's health can be ascertained as soon as possible, the hidden dangers can be discovered and a prompt diagnosis and treatment can be made. However, there are also cases like this: When a person gets a physical examination and finds that there is a small tumor in his body, or the cells show signs of cancer, he immediately starts surgery, chemotherapy, and has a heavy burden of thoughts, even a mental breakdown, which accelerates his death. Another person in the same situation, does not get a check-up, muddles through life, but enjoys the years of heaven. So it is said that cancer does not have to be detected as early as possible. Although this is one-sided, it also makes sense. I'm not a medical expert. It has been estimated that there are tens of billions of cells in the human body. There are always some cells that need to change and evolve towards cancer cells, but the inherent ability to "correct" that the human body possesses can often be controlled. The so-called "positive ability to suppress evil" (the correction ability of young people is stronger) in traditional Chinese medicine is not a problem. Just as there are always some people in the society who want to turn bad, as long as they are under the control of the people and the public security authorities, the "weather will not change". It can be seen that the problem is not whether to check the body, but what attitude is taken to treat the results.

The investigation and test work of a water conservation project before construction, to some extent, is like a physical examination. This work is always in-depth and step by step, and it is also like when people get a simple general physical examination first, and then receive the detailed specialty check-up. Through detailed investigation, more data will be obtained, which may be good information, and more new adverse factors will be found: a new fracture or a weathering zone is found thicker than expected, the strength of the material is reduced, etc. In addition, the in-depth analysis and calculation work will also expose some new situations, such as a great deal of stress and deformation, low safety and so on. It is often encountered that the safety design considered in the feasibility study fails to meet the requirements after the work is enlarged. For example, when checking the stability of the body of a dam or its slope, because more small joints are found, more tests are made, or a more accurate analysis is carried out, and it is found that the degree of safety does not meet the requirements of the specification. How can this problem be treated? I think if the new data does not change the large pattern, only the adjustment of specific data, and the survey and analysis means and depth have exceeded the conventional requirements, then it is questionable whether the design should be changed to a great extent. In fact, there is no strict scientific solution to the problems of accounting for the stability of a slope, and the design depends on experience and judgment to a certain extent. In other words, the means of exploration are matched to the depth, the test and the value of the parameters, the theory and method of calculation, and the required degree of safety. The safety factor required in the code is precisely specified considering the current level of exploration, testing and calculations. If we take particularly advanced measures or do work that is too detailed in order to find out very clearly what the problem is, we will not need such a high degree of security. Otherwise, there will be a contradiction that the deeper the work goes into the building, the less safe it will be. For the same reason, some comrades have developed a new theory of calculation for some structures (such as soil-rock dams) and they have obtained a greater degree of safety. I also do not like the achievement of modifying the design and reducing the quantities. In a word, it is absolutely necessary to develop new and more accurate means of surveying, testing and analysis, but it needs to be studied at a higher level if we want to modify the original matching of the design principles and specifications.

Some comrades have also made a choice according to the data at that time when choosing the dam, and regretted finding that the chosen dam site had some shortcomings after further work. In my opinion, as long as the new data does not negate the original main conclusion, there is no need to be upset. For another dam site, new problems will also be found after further work, but the dam sites with different survey depths cannot be compared. Just like when comparing a person who has only had a simple medical examination and a person who has done a comprehensive and detailed one, it is difficult to comment on who has more problems from the results of the examination.

**The Philosophy of Managing Children**

If we don't follow the rules, we cannot make a square. So we have to establish rules, standards and laws, from design and drawing to governing the country. If children are allowed to go astray, they will degenerate easily, so they should be disciplined. This is known to all. However, if you have to use rules to draw and write, there will be no calligrapher or painter. If there are many laws in a country, the common people do not know what to do, and they will be exploited by bad people, "the country will be ruined". It is also a common fact that if we are too strict with our children, they will have a reverse psychology. It can be seen that everything has a "degree", and too much is as bad as too little. The characteristic of hydraulic structures is that their failure not only means damage to the project itself, but it also causes great loss lives and property. In particular, the accident of a high dam and a large reservoir even means the destruction of the ecology and society, which is incomparable with other buildings. Therefore, governments of all countries need to manage their design and construction to different degrees.

In the past, I was deeply interested in the formulation of various codes, specifications and standards for water conservation and hydro power projects, especially when I was the chief engineer of the Ministry of water and electricity. Under my chairmanship and organization, I cleaned up and revised the issued norms, made plans for the absence of them, organized forces, and started the preparation in an all-round way, hoping to be exhaustive. It is believed that in this way, comprehensive control can be carried out and the world will be peaceful. China's system of norms and standards is based on those of the Soviet Union, which has been too trivial. In this way, they are "even more excellent than the previous ones". China has a great number of specifications for water conservation projects, probably many more than any other country.

Now I've been thinking about the consequences. Positive effects do exist, for example, in favor of the work of grassroots comrades, and in favor of ensuring the safety of small projects, but the negative effects are also great. First of all, innovation is hindered and a shackle is put on scientific and technological personnel. Norms are based on the past experience and have the responsibility to guide and restrict the national water conservation construction work. Therefore, we must recommend and stipulate relatively mature practices. It is good to achieve "above intermediate safety", and it is impossible to recommend the latest and most innovative methods. The norms are full of taboos such as "forbidden", "not allowed to" and "should not", and orders such as "must", "should" and "comply" exist. Some new ideas, theories, methods and techniques are all rejected under the cap of violating the norms. Second, norms provide an umbrella for those who do not want to make progress. Since so many regulations have been listed, and standard theories, methods, parameters and even specific formulas have been provided, it is not only safe and labor-saving to draw the ladle according to the sample, and all kinds of inspection, appraisal and acceptance cannot affect me; if something goes wrong, the specification can also be used as a defense weapon. If you make any innovation, you are responsible for everything. Therefore, the scale of China's water conservation construction work in recent decades is very large, which is rarely seen in the world. Although much scientific and technological progress has been made, the innovative achievements are not commensurate with it. The strength and speed of innovation cannot catch up with those of developed countries, and the excessive and detailed specifications may be one of the reasons. How can you expect your child to be Edison or Einstein if you set so many rules and regulations for him to treat or listen to or do anything without abiding by those rules and regulations?

So my idea now is that the standard should be less and vaguer, and the manual should be more and more detailed. Necessary norms and standards still need to exist, which should be in the position of "middle front". It is mainly required to specify some basic requirements and responsibilities of personnel at all levels, as well as approval procedures of projects at all levels. It is more important and effective to implement responsibility (including safety and advancement) to specific personnel than to set out indicators, parameters and methods. Some may be concerned that this will lead to a decline in the quality of the project and an increase in the number of accidents. In fact, there are many countries that do not have many norms, and even fewer are issued by the government. Most of them are standards and regulations formulated by some authoritative academic groups. They are followed by everyone and have no legal effect. As long as you have enough bases, you can break through. As a result, new things come out quickly, and no major accidents have happened.

As for the specific building structure, there is no need to make unified provisions for the specific formula for calculations, which can be changed into a large number of Design and Construction Manuals. The theories, various experiences, parameters and various typical examples can be collected and compared for the reference of the majority of grassroots personnel, rather than "adopting the same rule".

It is an art to make children full of vitality, who can keep innovating and moving forward, and not let them rush around like wild horses. Relevant government officials, science and technology managers and technology leaders should learn this art.

**The Philosophy of Eating Crabs**

As mentioned above, too many and too detailed specifications restrict innovation, which involves the use of new technologies in water conservation projects. The so-called new technology includes new ideas, new theories, new structures, new materials, new equipment, new processes, new management methods, etc. Technological innovation is of great significance, but the problems are also complex.

Any new technology, called "new", means that there is a lack of practical experience and it brings certain risks with it. The success or failure of water conservation projects usually has a huge impact. How to ensure the safety and promote the application of new technology has become a pair of contradictions. Figuratively speaking, "it is the one who will eat the first crab".

In my opinion, there are two correct ways to solve this contradiction: First, in terms of policy, we should develop policies of "prudence" and "positivity" at the same time. So-called "prudence" means keeping a clear mind and not doing things without a foundation or considerable assurance. Everything is decided by experiment and practice. Especially in important projects or key parts, we should not take things lightly. So-called "positive" refers to the ideological conviction that innovation and development are the right way of the world, the enthusiastic welcome of new things, in-depth investigation and research, the adoption of various measures to create conditions for their maturity and adoption, and the "dare to be the first" mind, rather than waiting passively and "eating after others". Second, in terms of strategy, the words "seek truth from facts and deal with each case on its merits" are extremely important. From conception, germination to maturity and promotion, new things generally go through three stages: theoretical research and laboratory testing, intermediate testing or industrial testing, and comprehensive promotion and productivity formation, but different methods and speeds can be adopted according to different situations.

Some new things have a solid, concise and credible theoretical basis, sufficient manpower, material resources, funds and time invested in research, and complete and systematic research and test results, while others fail to meet these requirements; the former can obviously be tried out and popularized on a faster and larger scale.

Some innovations are technological improvements, simple and clear in principle, easy to test and implement; others are major reforms in nature, with far-reaching consequences, and can only be determined after several years (such as some new materials). Obviously, the former can be tried out and popularized on a faster and larger scale.

If a new technology is adopted in some projects, the risk and consequence are not great, or there are conditions for repair and replacement; if a new technology is adopted in some projects or buildings in case of failure, the consequence is serious, or there are no conditions for repair and reinforcement, and it is obvious that the former can be bolder. Some of the successful new technologies in small projects need to be tested in medium-sized projects and then extended to large projects, but some can also be expanded by leaps and bounds through theoretical analysis and judgment.

Some engineering facilities leave more room (for example, there are more flood discharge holes, discharge holes, more units, etc.), which creates good conditions for the trial use of new technology. When conditions are right, engineers can also reserve special parts for testing new technology in engineering design. If we take this initiative and positive attitude, we will greatly promote the application of high-technology in water conservation projects.

**(Reprinted and translated from the article in Chinese, published the Journal of the China Institute of Water Resources and Hydropower Research, June 2003, Issue 1, Volume 1)**

## 7.2 Dialectics of Reform and Opening-Up—Learning from General Secretary Xi Jinping's Important Exposition on the Methodology of Comprehensively Deepening the Reform

**Feng Jun**

This year marks the 40th anniversary of the reform and opening-up. In the past 40 years, China has made remarkable historical achievements for many reasons. One of the important reasons is that the Communist Party of China has always adhered to the guidance of dialectical materialism and historical materialism in the historical process of leading China's reform and opening-up, and has consciously applied the Marxist world outlook and methodology. Since the 18th National Congress of the Communist Party of China, General Secretary Xi Jinping has led the Politburo to collectively learn the basic principles and methodology of dialectical materialism, the basic principles and methodology of historical materialism, and has constantly enhanced the capacity of dialectical thinking and strategic thinking, and has striven to improve the ability to solve the basic problems of China's reform and development. General Secretary Xi Jinping's important exposition on the methodology of comprehensively deepening reform is an important guiding ideology for comprehensively deepening the reform, a dialectics of reform and opening-up, a deepening and development of the basic principles of dialectical materialism and historical materialism, and an important part of the latest achievements of Sinicization of the Marxist doctrine.

1. **Correctly understand the relationship between emancipating the mind and seeking truth from facts, handle the relationship between being able to change and not being able to change, and the relationship between change and no change in the reform and opening-up**

The reform and opening-up started from the emancipation of the mind. The discussion on the standard of truth is a great ideological movement of emancipation, which became the ideological forerunner of the Third Plenary Session of the 18th Central Committee of the Communist Party of China and opened the curtain for the reform and opening-up. Without emancipating the mind, it is impossible for our party to make a historic decision to transfer the Party and the center of national work to economic construction, implement reform and opening-up, and embark on the road of socialism with Chinese characteristics.

Emancipating the mind and seeking truth from facts are closely related and complementary, and they are essentially the same. First of all, the starting point for emancipating the mind is to proceed from reality rather than from subjective conjecture, inherent experience and the book. The goal of emancipating the mind is to conform subjectively to the objective, theory to the reality, solve practical problems, understand the world and transform the world. Comrade Deng Xiaoping said, "To emancipate the mind is to make it conform to reality, to make it subjective and objective, and to seek truth from facts." Comrade Deng Xiaoping closely associated the meaning of "emancipating the mind" and with that of "seeking truth from facts". After the 14th National Congress of the Communist Party of China, "emancipating the mind and seeking truth from facts" was summarized as the Party's ideological line.

Today, we still need to "further emancipate the mind, further emancipate and develop social productivity, further emancipate and enhance social vitality" in order to comprehensively deepen the reform. If we do not further emancipate our minds, our party will not be able to continuously promote theoretical and practical innovation in practice, cope with various risks and challenges on the way forward, push reform and opening-up forward, and always walk in the forefront of the times. General Secretary Xi Jinping said that in this "three further emancipations", "emancipating the mind is a prerequisite, a general switch to emancipate and develop social productivity, emancipate and enhance social vitality". In "emancipating and developing social productivity, emancipating and enhancing social vitality is the inevitable result of emancipating the mind, and also an important basis for emancipating the mind".

However, among the "three further emancipations", further emancipation and development of social productivity is the most fundamental and urgent task and the ultimate goal. "Emancipating the mind, emancipating and enhancing social vitality is to better emancipate and develop social productivity".

On the basis of a correct understanding of the relationship between emancipating the mind and seeking truth from facts, we should deal with the relationship that can and cannot be changed in the reform and opening-up. At the beginning of reform and opening-up, Comrade Deng Xiaoping pointed out: "What is emancipation of the mind? When we talk about emancipating the mind, we mean breaking the shackles of habitual forces and subjective prejudices, studying new situations and solving new problems under the guidance of Marxism. Emancipation of the mind must not deviate from the track of the four cardinal principles or impair the political situation of stability, unity and liveliness." Today, the purpose of deepening the reform in an all-round way is to uphold and improve the Party's leadership and the socialist system with Chinese characteristics, from which we cannot deviate. Some hostile forces and people with ulterior motives define reform as a change in the direction of the Western political system, otherwise it will be regarded as non-reform. General Secretary Xi Jinping pointed out in a specific way: "Our reform and opening-up is a direction, a stand and a principle". "Our reform is a continuous reform on the road towards socialism with Chinese characteristics. It does not take the old road of blocking and ossification, nor does it change the evil way to change its flag". "The essence of the problem is what to change and what not to change. Some things that cannot be changed will remain unchanged for a long time. We cannot call this non-reform. We are constantly pushing reform forward in order to promote the better development of the cause of the Party and the people, not to cater to the'applause' of some people. We cannot mechanically copy Western theories and views onto our own. We should proceed from China's national conditions and from the reality of economic and social development, step by step to promote reform, without seeking sensationalism or superficial writing, and always adhere to the right direction of reform and opening-up." Reform and opening-up involves the issue of what banner to hold and what path to take. We should keep our heads clear, stick to the right direction, deal with the relationship between what can be changed and what cannot be changed, resolutely change what should be changed and what cannot be changed, resolutely hold on to what should not be changed, resolutely not change what cannot be changed, maintain the determination of reform, and never make subversive mistakes. We should proceed from China's national conditions, from the reality of economic and social development, and carry out reform step by step with leadership, which is the attitude of seeking truth from facts.

On the basis of a correct understanding of the relationship between emancipating the mind and seeking truth from facts, we should also deal with the relationship between changes and no changes in the judgment of China's basic national conditions in the reform and opening-up. After 40 years of reform and opening-up, the main contradiction in our society is no longer the contradiction between the people's growing material and cultural needs and backward social production. The report of the 19th National Congress of the Communist Party of China points out that "socialism with Chinese characteristics has entered a new era, and the main contradiction in our society has been transformed into the contradiction between the people's growing needs for a better life and the unbalanced and inadequate development." When we see the change, we also see that there is no change. "The change of the main social contradictions in our country has not changed our judgment on the historical stage of socialism in our country. The basic national conditions of our country, which is still in and will be in the primary stage of socialism for a long time, have not changed. The international status of our country, which is the largest developing country in the world, has not changed."

In the same way, we should not only see that the world is in an era of great change that has not happened in a century, but we should also see that China is in the best period of development since the beginning of modern times. We should not only see that China's economic development has entered a new normal, but we should also see that when facing international and domestic risks, we still need to make steady progress and maintain our strategic focus. To accurately grasp the new changes and new characteristics of the different stages of development in China, to make the subjective world conform to the objective reality better, and to determine the working policy according to the actual situation, is to adhere to the ideological line and scientific methodology of seeking truth from facts.

2. **Learn and master the basic principles of the contradictory movement of things, and deal with the relationship between the overall promotion of comprehensively deepening the reform and making key breakthroughs**

General Secretary Xi Jinping pointed out: "We must learn to grasp the basic principles of the contradictory movement of things, constantly strengthen the sense of problem, and actively face and resolve the contradictions encountered in the process. The problem is the manifestation of the contradiction of things. We emphasize enhancing the awareness of the problem and adhering to the orientation of the problem, that is, recognizing the universality and objectivity of the contradiction, that is, being good at understanding and resolving the contradiction as a breakthrough to open up the situation for work."

Contradiction is universal and objective. It is the right attitude not to avoid contradictions, dare to fight and be good at solving them. In his report at the 19th CPC National Congress, General Secretary Xi Jinping said, "Society is moving forward in the movement of contradictions, and there will be struggles when there are contradictions. In order to unite and lead the people to effectively cope with major challenges, resist major risks, overcome major resistance and solve major contradictions, our party must carry out a great struggle with many new historical characteristics. Any thoughts and actions that are greedy for enjoyment, passive and lazy, and avoid contradictions are wrong." In order to realize the great dream, we must have a great struggle, for example, to fight against all words and deeds that weaken, distort and negate the leadership of the Party and the socialist system; to fight against all acts that harm the interests of the people and are divorced from the masses; to fight against all stubborn and chronic diseases in the reform; to fight against all acts that split the motherland, undermine national unity and social harmony and stability. We should fight against all difficulties and challenges in the fields of politics, economy, culture, society and nature. The correct attitude to the contradiction should be to face the contradiction directly, and to promote the development of things in the process of solving the contradiction by using the characteristics of contradiction which complement each other.

The law of the unity of opposites of materialistic dialectics is usually called the "Two-Point Theory", that is, we should see two aspects of contradiction when we look at problems and do things, and we should see the struggle, balance, interdependence and transformation between the positive and the negative. If we do this, we will adhere to the "Two-Point Theory". When we look at problems and do things, we will not be simple, one-sided and rigid. We should adhere to the "Two-Point Theory" as well as the "Key-Point Theory". One thing is often a collection of contradictions. Among many contradictions, one of them must be the main contradiction. The main contradiction determines the nature and the direction of the development of things. If we grasp the main contradiction, other contradictions will be solved. In a contradiction, the two aspects of the contradiction are not completely equal and balanced. The main aspects of the contradiction determine the nature and the direction of the development of the contradiction. We must pay attention to the main aspects of the contradiction and pay attention to the conditions under which the main and secondary aspects of the contradiction will transform each other. To grasp the main contradiction in many contradictions and the main aspects in one contradiction is to adhere to the "Key-Point Theory".

General Secretary Xi Jinping proposed that priority should be given to resolving the main aspects of major contradictions and contradictions so as to drive other contradictions to be resolved. For example, the main contradictions that must be solved in the current development of the Party and the state are to coordinate and promote the building of a moderately prosperous society in all respects, to comprehensively deepen the reform, to advance the law-based governance of China and to strengthen Party self-conduct in an all-round way. General Secretary Xi Jinping also stressed the need to learn and grasp the basic method of the analysis of the social contradictions, and thoroughly understand the importance and urgency of deepening the reform in an all-round way. Only by combining the contradictory movements of productive forces and relations of production with those of economic foundation and superstructure, and observing the basic social contradictions as a whole, can we fully grasp the basic outlook and direction of the development of the entire society.

To fulfill the historic mission of the Party in the new era of socialism with Chinese characteristics, General Secretary Xi Jinping pointed out: "The decisive role in the great struggle, great project, great cause and great dream is the new great project of building up the party in the new era." To grasp the great project is to grasp the main contradiction. General Secretary Xi Jinping also put forward the idea of "two revolutions", one is social revolution, the other is self-revolution, and the socialist revolution of adhering to and developing socialism with Chinese characteristics is a social revolution. "To carry out the great social revolution of upholding and developing socialism with Chinese characteristics in the new era, our party must have the courage to carry out self-revolution and build up the Party, making it stronger and stronger." The Party's self-revolution is the most important. Only if the Party's self-revolution is done well can we better promote the Party to lead the people to carry out the great social revolution.

In the process of deepening the reform in an all-round way, the task of reform is heavy and the quantity is great and widespread. It needs to be promoted as a whole and implemented in an all-round way. However, we must pay attention to key breakthroughs, highlight key points, and make a comprehensive plan. There are many complex fields for political, economic, cultural, social, ecological civilization and party-building. The reform of the economic system is the key to comprehensively deepening the reform. "In comprehensively deepening the reform, we should adhere to the reform of the economic system as the main axis, strive to make new breakthroughs in the reform of important fields and key links, so as to lead and drive the reform forward in other fields."

Deepening the reform in an all-round way is the unity of overall promotion and key breakthroughs. We should adhere to the overall promotion and "learn to play the piano with ten fingers". The overall promotion is not to make equal efforts and keep pace with each other, but to pay attention to the main aspects of major contradictions and contradictions, and to the important fields and key links, which adheres to the unity of the "Two-Point Theory" and the "Key-Point Theory".

3. **Learn and master the principle of the dialectical relationship between understanding and practice, and deal with the relationship between top-level design in reform and opening-up and "crossing the river by feeling the stones"**

General Secretary Xi Jinping pointed out: "We should learn to grasp the principles of understanding and practicing dialectical relations, adhere to the viewpoint of "practice first", and constantly push forward theoretical innovation based on practice. We need to practice to produce real knowledge in order to advance our work. Theory must be unified with practice."

The process of reform and opening-up is also a process of constantly exploring and absorbing the wisdom of the people. "Crossing the river by feeling the stones" is a reform method full of Chinese wisdom, which is not only in line with China's national conditions, but it is also in line with Marxist epistemology and practice. In the past 40 years, many of China's reforms have been created by the people themselves. First, we should practice and work locally, pilot and explore, ask for the way, and then summarize our experience, so as to promote policies and systems across the country. Comrade Deng Xiaoping said in a talk in South China, "In rural areas, the right of invention belongs to the peasants, who are engaged in household contract for joint production. A lot of things in rural reform are created at the grass-roots level. We take it as the guidance of the whole country for processing and improvement." General Secretary Xi Jinping said, "China's reform and opening-up has come about this way. It is the process of first trial, then summary, and further promotion and continuous accumulation. It is the process of expanding from rural areas to cities, from coastal areas to the mainland, from local to the whole. This gradual reform avoids the social turbulence caused by an unclear situation and improper handling, and provides a guarantee for steadily advancing the reform and achieving the goal smoothly. It is in line with the process of people's understanding of objective laws and the dialectics of things from quantitative change to qualitative change."

In the early stage of reform and opening-up, our main method is to "cross the river by feeling the stones". With the deepening of the reform and opening-up, the accumulation of experience accumulation is more and more abundant, the understanding of the law of reform is more and more profound, and the top-level design is more and more important. In the top-level design, we need to have systematic thinking and overall thinking, and we need comprehensive balance and overall consideration. "As for comprehensively deepening the reform, 'comprehensive' means promoting the reform in all fields as a whole; we need to have an overall goal, and also answer the question of what is the ultimate goal of promoting the reform in all fields and what kind of overall results we want to achieve." To do a good job in overall planning, we should not only " thoroughly study the top-level design and overall planning of comprehensively deepening the reform of the system, clearly put forward the overall plan, road map and timetable of the reform", but also comprehensively promote the "five-sphere integrated plan", make an overall design and specific arrangements for the economic system, political system, cultural system, social system and the system of an ecological civilization, as well as vigorously promote the reform of scientific decision-making. Deepening the reform in an all-round way is a major deployment related to the overall development of the Party and the state, not a single reform in a certain field or aspect. We should start from the overall situation, have a long-term vision, really look forward, think ahead and plan ahead. We should "correctly handle the relationship between the central and local governments, the overall and local governments, the long-term and current ones, correctly handle the adjustment of the pattern of interests, and resolutely overcome the constraints of local and departmental interests".

Today's reform still "needs to take the method of pilot exploration and "throwing a stone to clear the road". We should push after having gained experience, formed a consensus, seen very accurately, and felt that the push is very stable, so as to accumulate small wins to achieve big wins." "To cross the river by feeling the stones" is not to feel aimlessly, "There are rules for crossing the river by feeling the stones. We should follow the recognized rules, and deepen our understanding of the rules in practice, instead of stepping on the watermelon skin and not knowing where to go." China is a big country with a vast geographical area, a large population, a great difference in nationality and geography, and a great difference in economic development, culture, education and science and technology. When we think about problems and make decisions, we should start from the local conditions, take measures according to those local conditions, not be reckless, not be uniform and one size fits all, and never make subversive mistakes on fundamental issues. "Crossing the river by feeling the stones" is a reform method that we have explored that is in line with China's national conditions. Therefore, "crossing the river by feeling the stones and strengthening the top-level design are dialectically unified. The promotion of a partial staged reform and opening-up should be carried out on the premise of strengthening the top-level design, and the strengthening of the top-level design should be planned on the basis of promoting a partial staged reform and opening-up".

4. **Learn to grasp the relationship between respecting objective laws and exerting subjective initiative, and deal with the relationship between boldness and stability in reform and opening-up**

General Secretary Xi Jinping pointed out: "We must handle the relationship between respecting objective laws and giving full play to subjective initiative well. We should adhere to the principle of starting from reality and following the objective laws. We should follow a blueprint to the end and lay a solid foundation for the long-term work. At the same time, we should encourage the local, grass-roots and the masses to explore boldly, to try first, to advance innovation in theory and practice, and to deepen our understanding of the law of reform."

In 1992, Comrade Deng Xiaoping said in a talk in South China, "We should be bold in reform and opening-up, dare to experiment, and not be like little women. If you are sure of it, you should try boldly and break through boldly." Without the spirit of "pioneering" and "taking risks", no new career can be achieved. Comrade Deng Xiaoping also hopes that regions with optimal conditions can move forward faster economically. China's reform and opening-up is a gradual reform, prudent, and seeking progress in stability is the basic strategy of China's reform and opening-up. Some countries have engaged in so-called "shock therapy", which has resulted in severe social unrest. The lesson is profound. Before the promulgation of our reform policies and measures, we have repeatedly demonstrated and scientifically evaluated them, and gradually promoted them from point to surface, maintaining policy continuity and stability. Therefore, to deal with the relationship between courage and stability has always been one of the important methodologies of reform and opening-up.

General Secretary Xi Jinping discussed the dialectical relationship between courage and stability. He said, "We should have the courage, which means that no matter how big the difficulty is, we have to move forward, be bold in taking responsibilities, be bold in 'gnawing at hard bones', and be bold in wading through dangerous shoals. A steady step means that the direction must be accurate and the driving must be stable, especially that no subversive mistakes should be made. We should be bold in pushing reform forward, but we must make steady progress. Courage is not recklessness. Recklessness will lead to blind tossing. For some major reforms, it is impossible to finish the first battle, and we can put forward overall ideas and plans, but we still need to work steadily to achieve the goal step by step through continuous efforts, and accumulate small victories in order to achieve great victories. This is called 'It's hard to make a picture easier than it is to make it finer. All the difficult things in the world must be done easily; all important things in the world must be done in detail.'"

Faced with the objective existence of nature and society and its objective laws, people are always playing their subjective initiative to understand and transform the world. Reform and opening-up is a new creation, emancipating the mind, taking the initiative, daring to do something and being bold, which is to give full play to the human's subjective initiative; but reform and opening-up must also respect the objective laws and act according to them. "Emancipating the mind is not a whimsical idea divorced from national conditions, a subjective imagination of building a car behind closed doors, or a reckless act without rules and regulations". To give full play to subjective initiative is not the subjective idealism of "how bold people are and how productive they are". Reform and opening-up should not only give full play to people's subjective initiatives, but also fully respect the objective laws, so that we can walk steadily.

5. **Learn and master systematic thinking, deal with the relationship between reform, development and stability, and pay more attention to the systematicness, integrity and coordination of reform**

General Secretary Xi Jinping stressed: "We put forward a plan for comprehensively deepening the reform because we need to solve the outstanding contradictions and problems that we face. We can only rely on the reform of a single field and a single level. We must strengthen top-level design, overall planning, and enhance the relevance, systematicness and synergy of all reforms."

Reform and opening-up and the comprehensive deepening of the reform are a huge systematic project, involving numerous factors and variables, which often lead to one development and move the whole body. Comprehensive deepening of reform should adhere to the overall and local matching, the combination of root and standard, and the promotion of gradual progress and breakthrough. Therefore, we should use thinking via a system to grasp such a huge system as comprehensively deepening the reform.

Reform, development and stability have a close and complex relationship. First of all, reform is for development. "Only by closely focusing on the first priority of development to deploy reforms in all aspects, and by liberating and developing social productivity to provide a strong traction for reform, can we better promote the adaptation of production relations to productivity, superstructure to economic foundation." Second, development promotes reform. Many problems exist only when they are solved during development. Only when a lot of reform measures are developed can they be introduced. Only when they are developed can they be changed. Third, stability provides a stable and harmonious environment for reform and development. Without a stable situation, neither reform nor development can be achieved. Conversely, without reform and development, people's lives cannot be improved, people's sense of gain cannot be achieved, and it is difficult to stabilize society. Therefore, "We should adhere to the unity of the strength of reform, the speed of development and the level of social affordability, regard improving people's lives as the junction of correctly handling the relationship between reform, development and stability, promote reform and development in maintaining social stability, and promote social stability through reform and development".

To deal with the relationship between reform, development and stability, we must enhance the systematicness, integrity and coordination of reform. First, we should unswervingly push forward reform and opening-up, with no pause in reform and no stop in opening-up. We should use greater political courage and wisdom to push forward the next reform. The deepening of reform in an all-round way touches on the adjustment of deep-seated social relations and interests, inevitably touching some people's "cheese", breaking through the vested interests, and not all of them will be happy. To implement the reform measures, we need to have courage and responsibility. Adhere to the correct direction of reform, dare to "bite the hard bone", dare to "wade through dangerous shoals", dare to break through the obstacles of ideas, dare to break through the barriers of interest solidification. Second, to build consensus on reform, it is difficult to form a joint force of reform and opening-up without a broad consensus. "Now, the economic system is undergoing profound changes, the social structure is undergoing profound changes, the pattern of interests is undergoing profound adjustments, ideas are undergoing profound changes, so it is more difficult to gather consensus on reform, and the task of balancing the interests of all parties is more and more arduous. This requires more efforts to build a consensus. It is very important to build a consensus. When the ideas are not unified, we need to find the greatest common divisor; if you find the largest common divisor, focus on reform and opening-up, you can do half the work and double the work." We should unite all the forces that can be united, mobilize all the positive factors that can be mobilized, and combine them into a strong force to promote reform and opening-up. Third, we need to implement policies in a scientific way and not blindly promote reform. We need to push forward in accordance with the requirements of the Central Committee and not beyond the limits set by the Central Committee; we need to advance reform in an orderly manner, without any rush or delay; we should not push forward the pilot in a hurry or rush to success. Fourth, to advance in a coordinated way, each reform will have an important impact on other reforms, and each reform needs the coordination of other reforms. We should pay attention to the relevance and coupling of the reform, avoid being too light or too heavy, and avoid each other's actions and constraints. "With the deepening of the reform and opening-up, the relevance and interaction of that reform and opening-up have been significantly enhanced, which requires us to pay more attention to the mutual promotion and benign interaction of various reforms". We should grasp the important relationship of comprehensively deepening the reform, make all reform measures cooperate with each other in policy orientation, promote each other in the process of implementation, and complement each other in the effectiveness of the reform, so as to form a strong force of reform and opening-up.

6. **Constantly enhance the ability of dialectical thinking, accurately grasp and properly handle various major relationships in different fields of comprehensively deepening the reform**

General Secretary Xi Jinping asked: "We must learn to grasp the fundamental method of materialist dialectics, constantly enhance the ability of dialectical thinking, and improve the ability to handle complex situations and deal with complex problems. The deeper our career is going, the more we need to constantly enhance our ability at dialectical thinking."

Comprehensively deepening the reform faces various difficulties and challenges, which requires us to be good at dealing with the relationship between local and overall, current and long-term, focus and non-focus, and to make the most favorable strategic choice in balancing the advantages and disadvantages. We should strengthen investigation and research, adhere to development rather than being static, comprehensive rather than one-sided, systematic rather than scattered, universal rather than single and isolated observation of things, accurately grasp the objective reality, truly grasp the law, and properly handle various major relationships.

The process of reform is also a process of coordinating and dealing with all kinds of relations. We should deal with all kinds of important relations in different fields of reform.

The general goal of comprehensively deepening the reform proposed by the Third Plenary Session of the 18th Central Committee of the Communist Party of China is composed of two sentences, namely, "improving and developing the socialist system with Chinese characteristics" and "promoting the modernization of the national governance system and governance capability". We need to correctly understand the relationship between the two sentences in the general goal of comprehensively deepening the reform, as well as the relationship between China's governance system and governance capability in the second sentence. "The previous sentence sets out the fundamental direction. Our direction is the road towards socialism with Chinese characteristics, not any other way. The latter sentence provides a clear direction for the improvement and development of the socialist system with Chinese characteristics under the guidance of the fundamental direction." These two sentences form a whole. Only when we put them together can we fully understand and grasp the overall goal of comprehensively deepening the reform. In the second sentence, we should also correctly understand and grasp the relationship between the national governance system and governance capacity. General Secretary Xi Jinping's exposition of the relationship between the system of state governance and governance capacity has three meanings. First, they have different functions, they complement each other and are indispensable; "they complement each other, neither country can be governed by itself. To govern a country, institutions play a fundamental, overall and long-term role. However, without an effective governance capacity, no good system can play a role". Second, they promote each other; "only with a good system of national governance can we improve the governance capacity, and only by improving the national governance capacity can we give full play to the effectiveness of the system of national governance ". Third, although the system of national governance and governance capacity are closely related, they are not the same thing and cannot be completely the same; "the more perfect the system of national governance system is, the stronger the national governance capacity will naturally be. Throughout the world, each country has its own system of governance, and there is a big or small gap in the governance capacity of each country due to the differences in objective conditions and subjective efforts, even the governance capacity of the same country in different historical periods under the same system of governance system has a big gap". Therefore, the system of national governance and governance capacity are not only different from each other, but are also complementary to each other. They are dialectical and unified.

Reform and opening-up are also a pair of relations, which should be handled well. In the process of reform and opening-up, only after the reform is carried out can it be opened up, and only when the reform is in depth can it be opened up in breadth. And opening-up has forced and promoted reform. Opening-up has broadened vision, broadened thinking, opened channels, and further promoted the deepening of the reform.

In the reform of the economic system, we should pay attention to the relationship among the government, the market and the society, so that the market can play a decisive role in the allocation of resources and play the role of the government better. We should deal with the relationship between large and small, between revenue and expenditure, between management and service, between public economy and non-public economy, and between economic development and ecological environmental protection. In the reform of the political system, we should pay attention to the unity of adhering to the leadership of the Party, the people as the masters of the country and governing the country according to law. In terms of developmental mode, we should pay attention to the relationship among innovative development, coordinated development, green development, open development and shared development.

Reform and opening-up is a great new revolution that our party leads the Chinese people in during the new era. It is the most distinctive feature of contemporary China and the most distinctive banner of our party. There is no end to reform and opening-up; it is always in progress and not completed. In order to deepen the reform in an all-round way and further the reform and opening-up, we must continue to accept the nourishment of the wisdom of Marxist philosophy, never leave the guidance of the Marxist world outlook and methodology, and never leave materialistic dialectics. We should consciously use materialistic dialectics to guide our actions, apply dialectics in the practice of reform and opening-up, and develop dialectics. Xi Jinping's socialist ideology with Chinese characteristics in the new era is shining in the glory of materialist dialectics. It is the crystallization of the collective wisdom of the Communist Party of China, the latest achievement of the Sinicization of the Marxist doctrine, the Marxist doctrine of contemporary China, and the Marxist doctrine of the 21st century.

(Author: Feng Jun, member and professor of the Central Academy of Party History and Literature)

For the original text, see *Guangming Daily*, November 19, 2018.

# References

1. Jun S (1996) Theory and practice of underground engineering design. Shanghai Science and Technology Press, Shanghai
2. Mengshu W (2010) China tunnel and underground engineering construction technology. People's Communications Press
3. Haifan X (2011) Bridge conceptual design. People's Communications Press
4. Qingyuan Z et al (2005) Theory and application of train derailment analysis. Central South University Press, Changsha
5. Hanhua Z et al (2012) Thinking on the safety of civil engineering structures under stress. People's Communications Press, Beijing
6. Hanhua Z et al (2004) Technology for prevention and control for the typical hazards of the Sichuan-Tibet highway in Tibet. People's Communications Press, Beijing
7. Compilation Committee of Geotechnical Engineering Manual (1994) Geotechnical engineering manual. China Construction Industry Press, Beijing
8. Compilation Committee of Engineering Geology Manual (1992) Engineering geology manual. China Construction Industry Press, Beijing
9. Institute of Mountain Hazards and Environment, CAS, Lanzhou Institute of Glaciology and Geocryology, Chinese Academy of Sciences, and Institute of Science, Department of Transportation of Tibet Autonomous Region (1995) Mountain Hazards in the South Line of the Sichuan-Tibet Highway (in Tibet) and Prevention and Control Measures. Science Press, Beijing
10. Institute of Mountain Hazards and Environment, CAS, and Institute of Science, Department of Transportation of the Tibet Autonomous Region (1999) Study on the Typical Mountain Hazards of the Sichuan-Tibet Highway. Chengdu University of Science and Technology Press, Chengdu
11. Shuming L (2015) East and West culture and philosophy. Shanghai People's Publishing House, Shanghai
12. Peilong Q (2009) 50 Differences in thinking between the oriental and the Westerner. Harbin Press, Harbin



H. Zhu and L. Shi, *Methodology of Highway Engineering Structural Design and Construction*, Advanced Topics in Science and Technology in China 59,
https://doi.org/10.1007/978-981-15-6544-1

The manufacturer's authorised representative in the EU is Springer Nature Customer Service Centre GmbH, Europaplatz 3, 69115 Heidelberg, Germany. If you have any concerns regarding our products, please contact ProductSafety@springernature.com

Printed and bound by CPI Group (UK) Ltd, Croydon, CR0 4YY

30/09/2026

02225333-0009